Earth is in the middle of a revoluti lead us to more knowledge about ourselves, our place in the universe and the universe itself, or it might lead us to catastrophe. This book is a report on seventy-five years of history during which extraterrestrials (ETs) and their extraterrestrial vehicles (ETVs;, which used to be called UFOs) – have made themselves a part of life on earth.

Don Donderi, psychology professor at McGill University in Montreal (retired), tells us in *Truth, Lies and ETs: How we Stumbled into the Universe*, that Alien visitors have been kidnapping humans and treating us like specimens for at least 75 years. Beginning during the last years of World War 2, visitors from elsewhere in the Universe began arriving in their "flying saucers." They haven't left, and have been observing and experimenting with us ever since. The uninvited extraterrestrial visitors (ETs) kidnap and examine people; interfere with human reproduction to create hybrid human-aliens, and come and go as they please.

While every part of this story has been told before by careful and dedicated researchers, Donderi assembles it all into one easily accessible volume that makes it hard to ignore: what is really happening (Truth), how governments have tried to conceal what is happening (Lies), and who is doing this to us (ETs).

His book is a concise summary of our current uncertain status on the planet that we used to think was ours alone. It brings together facts that, taken one at a time, might easily be dismissed, forgotten or ignored. But when the simultaneous impact of all the facts are considered, one conclusion is inescapable, that we humans are no longer masters of our planet. Extraterrestrials treat us the way we routinely treat earth's lesser species: tracking, capturing, and breeding them and keeping them under surveillance. We have to realize that this is happening before we can act together to regain control of our destiny on earth.

Praise for Truth, Lies, and ETs

"Donderi presents a daring and scary hypothesis, but his research is so painstaking, his argument so compelling - and besides, UFOs (now called ETVs) are a fascinating subject - so I would recommend this book to open-minded readers so they can join in what will soon become a major discussion."— *Susan Jean Palmer, Affiliate Professor, Department of Religions and Cultures, Concordia University Montreal*

"Retired psychology professor Don Donderi's book is "Spare, cogent, forceful, and direct. Even inveterate doubters will find themselves questioning their own assumptions, their job being, as Donderi writes, 'to find alternative explanations for the evidence.'"—*Keith Henderson, author of Mont Babel*

"If you want to know the truth about Extra Terrestrial Vehicles, read Truth, Lies, and ETs. Dr. Donderi presents the facts, the reports, and the hard evidence that other research scientists ignore and professional disinformants deny. A highly recommended page turner."— *Kathleen Marden, ETV and abduction researcher, on-camera expert and author, with the late Stanton Friedman, of Captured: The Betty and Barney Hill UFO Experience.*

Truth, Lies and ETs:

How We Stumbled into the Universe

Don Donderi Ph.D.

Moonshine Cove Publishing, LLC

Abbeville, South Carolina U.S.A.

First Moonshine Cove Edition March 2022

ISBN: 9781952439285

Library of Congress LCCN: 2022903629

Dedicated to the memory of

Verna Holcombe Donderi

1937 - 2020

About the Author

Don C. Donderi is a retired McGill University psychology professor and the co-founder and principal consultant of a Toronto-based human factors and ergonomics consulting firm. Donderi entered the University of Chicago in 1953 at age 15 and graduated with a BA in 1955 and a BSc in biological psychology in 1957. He then attended Cornell University where he graduated with a PhD in experimental psychology in 1963. While at Cornell, he worked for IBM's Federal Systems Division, designing radar navigation displays for the B-52 bomber. He has been studying the psychological aspects of the UFO and ET evidence since 1965. The author of two psychology textbooks and many research publications, he also wrote *UFOs, ETs and Alien Abductions: A Scientist Looks at the Evidence* (Hampton Roads Press, 2013). He is a dual American-Canadian citizen and lives in Montreal.

ufoets.com

Acknowledgements

Family

The book is dedicated to the memory of my wife and best friend, Verna Holcombe Donderi. I lived with Verna, a woman of wit, good sense and superb taste, from 1957 until she died in 2020. Our children, Andrea and Douglas, are the happy outcome of our long and happy marriage. Andrea and Johathan and Douglas and Tamara are the closer part of an extended family that includes children, nephews, nieces, cousins and in-laws who live across North America and around the globe, and I am happy to be part of this family. The book was begun at Douglas and Tamara's dining-room table in Toronto during the covid-19 pandemic. Andrea and Douglas read the manuscript and I profited from their suggestions.

Community

I first met members of the ET study community at David Jacobs' house in Wyndmoor, PA, in the early 'nineties. I met more of them at the late Budd Hopkins' house in New York City, and then through the Center for UFO studies in Chicago, directed at the time by Mark Rodeghier. I have participated recently with the Mutual UFO Network Experiencer Resource Team led by Kathleen Marden and George Medich. I appreciate their work, their online company and their continued commitment to understanding our ET problem. I thank Mark Rodeghier, Don Schmitt, George Eberhart and Patrick Huyghe for help with and advice on permissions and citations. I have enjoyed the company and hospitality of members of the ET study community in Montreal and the rest of Québec. They include Wido Hoville, Marc Leduc, Benoit Meilleur, Susan Palmer, Marc St-Germain, Alex Sazonov, Yann Vadnais and Luigi Vendittelli.

I thank Marc St-Germain for his help with Les Enfants de Sylvie P and Luigi for his help with the Harare, Zimbabwe case. Montrealer Keith Henderson has mastered the art of transforming fact into fiction and has written about some of the experiences that were described here (101). His fiction helps, as fiction does, to highlight the reality that it depicts.

In 2015 Verna and I enjoyed the hospitality of Younghae Chi, of Oxford University's Faculty of Oriental Studies, when he invited me to try to persuade Oxford undergraduates to pay attention to ETs. I remember with respect the life and work of my friend and university classmate, UFO investigator Stanton T. Friedman (1934-2019). Stanton, along with Kathleen Marden, wrote that "Science was wrong." Science was wrong for years.

The community of hundreds of people all over the world who study ETs and ETVs is too big to thank individually. I thank them all for the clear-sightedness and perseverance that has made this book possible. During the more than 50 years that I have been studying ETs and ETVs, the people with whom I have shared conversations, correspondence, discussions and talks have sustained my interest and made me realize that many of us know we have an ET problem. My hope is that Truth, Lies and ETs persuades everybody that we have an ET problem.

CONTENTS

TRUTH, LIES, AND ETS

Introduction

Earth is in the middle of a revolution. It might lead us to more knowledge about ourselves, our place in the universe and the universe itself, or it might lead us to catastrophe. I tell the story of this revolution as we have experienced it over the past seventy-five years. I explain where we are in the revolution and how we have managed or mismanaged it. I write about what might happen in the future and how we should prepare for what might happen.

The story of this revolution, already a part of our lives, is the work of hundreds of people who have experienced it, recorded it and analyzed it. They have had to overcome continued ignorance and resistance from people who do not believe what they have read or have been told about the revolution. This book is a report on seventy-five years of history during which extraterrestrials (ETs) and their extraterrestrial vehicles (ETVs; they used to be called UFOs) – have made themselves a part of life on earth.

The short version

You have read or heard something about ETs but may not know what to make of it. Here is a short explanation. ETs are intelligent beings from elsewhere in the universe who are curious about Earth, and curious about human beings because we are the dominant species on Earth. Extraterrestrial vehicles (ETVs) are the machines that carry ETs from other worlds to and from Earth.

ETs are telepathic, which means that they can communicate to us without talking. ETs can persuade us telepathically to do what *they* want us to do, instead of what *we* want to do. And their machines – ETVs – do things that our machines cannot do, but that we are slowly learning how to do.

ETs are more than just curious visitors: they are conducting breeding experiments that create human – ET hybrids.

ETVs (then called UFOs – unidentified flying objects) began to show up in large numbers in the mid nineteen-forties, towards the end of the Second World War. Some of the people who saw the UFOs founded private UFO study groups, and many people joined those groups. Much of what they learned and what their groups have published is presented in this book.

Other people worry that ETs might weaken or destroy human civilization. Some of them speak for governments or for government-funded organizations. They have tried to conceal, deny or discredit the ET evidence – while at the same time they continue to study it. Their goal might be to delay acceptance of ET reality until we have mastered ETV technology and can defend ourselves against telepathic control by ETs.

Other people ignore or deny the ET evidence because they are simply proud but ignorant professionals. Someone whose self-respect is based on his or her status as an expert may not be willing to admit that there is any part of the universe that is beyond the limits of what their education and training allow them to understand. Whether they are astronomers, astrophysicists, psychologists or have some other job that requires an advanced degree, these proud but ignorant "experts" dismiss or deny what they do not know and do not understand: they dismiss or deny the ET evidence.

That was the short version.

The longer version

This book tells the longer version of the ET story, which has three parts: *Truth, Lies,* and *ETs.*

Truth reports what we know about the ET visitors: their ETV technology, their telepathic abilities and what they are doing here. This story has gaps, and there are differences of opinion about details. The gaps and the differences are being studied by a community of people with an interest in the truth, and the gaps are being filled in and the details resolved as the evidence accumulates.

Lies is about the misleading statements and opinions that have been circulated, either by people whose interest is to obscure the truth about ETs or by people who just do not realize how little they know. The other side of that story – also told in *Lies* – is about the people who have worked and are working to document the truth and overcome the lies by collecting and reporting the ET evidence.

ETs reviews the challenges we face when dealing with ETs who have joined us on what we used to think of as our own planet. We have to decide what to do about ETs because we have no choice: ETs are here, and they come and go and interfere with us as they please.

About myself

Extraterrestrial vehicles (ETVs, then called UFOs) began to be reported in newspapers and magazines and to be discussed on the radio in 1947, when I was young and curious. Since then, I have acquired an education and some skills that are useful to someone interested in ETs and ETVs, and I am still curious. I am using that education and those skills to help describe and explain ETs and ETVs, and to propose how we should deal with their continuing presence on our planet.

I am a research scientist with a doctorate in experimental psychology. My research specialty is human visual perception and memory. My experience includes work in the aerospace industry as well as many years of teaching, research and administration at McGill University in Montreal. I co-founded a human factors and ergonomics consulting practice that serves military organizations, the aerospace industry, shipping companies, nuclear power plants, commercial businesses and the legal profession in North America and Europe. I have worked outside the ivory tower for as long as I lived inside it. When you work with pilots, nuclear engineers, ship's captains, businessmen and lawyers you learn that reality is more important than theory, and I have always preferred reality to theory.

The ET story begins with what people *see* and what they *remember* about what they see. Most of my professional life has been spent studying how people see and remember. I have been interested in the reality of what people see and remember about ETVs and ETs for most of my career. My approach to

ETs and ETVs is informed by my scientific training and by my work in applied science.

In 2013 I published *UFOs, ETs and Alien Abductions: A Scientist Looks at the Evidence* (1). There I defended the accuracy, reliability and importance of the ET evidence against the barriers to its acceptance resulting from various philosophical approaches to modern science. Much of the scientific establishment still ignores the ET evidence, and ill-informed 'scientific' opinions about the evidence circulate widely. *Truth, Lies and ETs* updates the ET evidence. It separates facts from the falsehoods, deliberate or misinformed, that have circulated about ETVs and ETs. The goal of this book is to help us understand and to meet the challenge that ETs present to the future of humanity on this planet.

Truth

Human experience, which is constantly contradicting theory, is the great test of truth.

Samuel Johnson

The ETV story changed dramatically on December 16, 2017, and it is still changing. I tell the new story in *ETVs – Yesterday.* I tell the older story in *UFOs – History*, which begins at the start of the ET era in 1944 and reviews the ETV evidence (then called the UFO evidence) from 1944 through 2017. *How do ETVs work?* describes some of the research and engineering developments that take us along the road to understanding ETV technology.

The *Extraterrestrials* story follows the ETV story. The first part of the story is about who ETs are. The second part, *Abductions*, is about what they are doing here -- abducting people into ETVs, studying them, interfering with their reproductive systems and then releasing them. *Who gets abducted and how?* explains some of the measurable differences between people who have been abducted, people who haven't been abducted, and people who pretend to have been abducted.

A Summary of the Truth concludes with a review of our interactions with ETs and a discussion of the problems that they are causing.

1. ETVs – Yesterday

Yesterday (relatively speaking, because the ETV story is now over seventy-five years old) the US government told us that ETVs were real. The *Washington Post* and the *New York Times* interviewed US Navy pilots about the ETVs they had seen, and videos of ETVs that had been recorded from their airplanes were released. This flood of new information arrived in late 2017 and early 2018, but it was about events that had been observed and recorded from 2004 through 2015. New ETV videos and stories continue to appear and they have now fully captured the attention of mainstream media (2).

These ETVs are not American, Chinese or Russian secret weapons: they fly faster and accelerate faster than anything we humans know how to make. They start and stop more abruptly than humans could survive unless we know how to counteract the deadly stress of rapid acceleration and deceleration – which we don't yet know how to do. The reports and videos were not the result of human error or failure of the electronic sensors aboard US ships and aircraft. The witness reports are numerous and the accuracy of the video evidence has been officially acknowledged. The U. S. Government's public response is: we don't know what's going on but we'll keep investigating (3).

This new information has changed what we know about our place in the universe. The next chapter, *UFOs – History,* tells us that it was not our first glimpse of that change. But the information in *ETVs – Yesterday*, coming from people who worked for the US Department of Defense and followed by confirmatory admissions from the US government, has put the older UFO evidence into clear focus. You can stop wondering whether UFOs are extraterrestrial vehicles and whether ETs exist. UFOs are extraterrestrial vehicles (ETVs) and ETs exist.

David Fravor and Alex Dietrich

Here are two of the recent stories that changed the world. The first story, told by US Navy Commander David Fravor and by Lt. Cmdr. Alex Dietrich (Figure 1), dates from 2004.

Commander Fravor was interviewed by the *New York Times* on December 16, 2017, about an encounter that happened thirteen years earlier. The video record of his encounter, and other similar video clips, were surreptitiously downloaded from the computer servers aboard the aircraft carrier *USS Nimitz* and then posted on an obscure website shortly after they had been recorded. In a reality-altering policy change by the US Department of Defense, three of the video clips – the original clip from Commander Fravor's flight, called "Flir" (Figure 2), and two later ones, called "Gimbal" and "Go Fast," were authenticated and "cleared for open publication" on August 24, 2017 (2).

Figure 1. David Fravor, USN and Alex Dietrich, USN

Commander Fravor commanded an F-18 squadron on the aircraft carrier *USS Nimitz*, which was cruising on the Pacific Ocean near the coast of California on November 14, 2004. The *Nimitz* was contacted by the cruiser *USS Princeton*, whose radar sensors had picked up multiple targets near the carrier. Fravor's jet was launched to investigate, along with an F-18 flown by Lt.-Cmdr. Dietrich. The objects seen on the *Princeton's* sensors moved from 80,000 feet to 20,000 feet in seconds. Fravor's sensors found one of the targets close to the ocean surface and he began to track and follow it. As he flew down towards it, it sped up and away to a point 40 miles distant. After hovering fifty feet above sea level, it had travelled at over three times the speed of sound to evade Fravor. Fravor said "I have no idea what I saw. It had no plumes, wings or rotors and out-ran our F-18s. I want to fly one." Dietrich saw the same thing from her F-18.

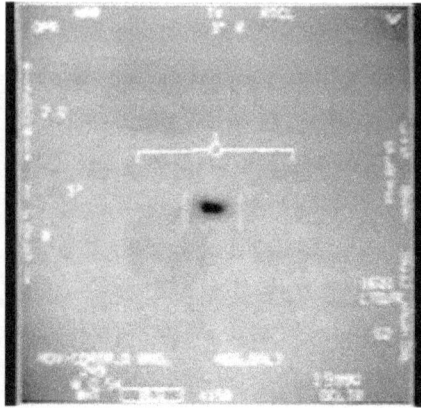

Figure 2. FLIR image from the sensor system in Commander David Fravor's F-18 Hornet

Commander Fravor and Lt.-Cmdr. Dietrich landed on the *Nimitz* and then then Lt. Cmdr. Chad Underwood's F-18 was catapulted into the air to follow up on Fravor's observation. Underwood saw the same thing that Fravor and Dietrich had seen. He radioed his observations back to the carrier. His observations, like Fravor's, were recorded, and the images from the radar and infra-red sensors aboard his aircraft were captured. He and Fravor called the ETV a "Tic-Tac." He said "Aircraft, whether they're manned or unmanned, still have to obey the laws of physics… the Tic-Tac was not doing that. It was going from like 50,000 feet to, you know, a hundred feet in like seconds, which is not possible"(4). After he landed, Underwood was called by NORAD (the North American Air Defense Command) and asked to describe what he saw. He did. They did not explain what he had seen, nor did they order him to keep silent about it.

Ryan Graves and Danny Accoin

The second story dates from 2014. Lieutenants Ryan Graves and Danny Accoin flew F-18 Hornet fighters from the USS *Theodore Roosevelt*, CVN 70. The carrier crew was preparing for combat missions in the Persian Gulf and they were training in the Atlantic Ocean off the Florida Coast. Graves and Accoin, both from the "Red Ripper" squadron, saw something "like a spinning top moving against the wind." "These things would be out there all day," Graves said (4). He said that the objects persisted, showing up at 30,000 feet, 20,000 feet even down to sea level. They would accelerate, slow down and then reach hypersonic speed. One pilot, watching a video of an object taken with the jet's gun camera, said, "Wow, what is that man? Look at it fly! (5).

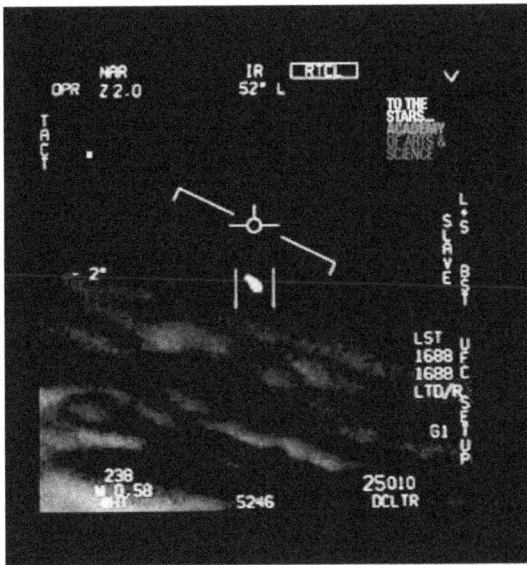

Figure 3: "Gimbal" Image from the USS Theodore Roosevelt.

This is what they saw (Figure 3). The image, like Commander

Fravor's, is from an infra-red imaging system. It was recorded from one of the F-18s based on the *Theodore Roosevelt* during the 2014 and 2015 sightings over the Atlantic Ocean, which took place off both the Florida and Virginia coasts (4). An artists' rendering of what was seen off both the Pacific and the Atlantic coasts of North America, and described by Lt. Cmdr. Woodward as a "giant tic-tac" is shown below (Figure 4):

Figure 4. Lt. Cmdr. Woodward's 'Tic-Tac' ETV

The US Government – Yesterday

In 2018, former US Deputy Undersecretary of Defense Christopher Mellon wrote an op-ed in the *Washington Post* that criticized the lack of coordination among the military services in studying the ETV problem, and urging the US government to publicly acknowledge its interest in ETVs. He wrote that "UFOs exist… So the issue now is: why are they here, where are they coming from and what is the technology behind these devices that we are observing?"(6). Good questions, all –answers will be suggested in the next few chapters.

The Pentagon's "advanced aerospace threat identification program" was first described in articles that appeared in the *New York Times* in 2017. The former manager of this program, Luis Elizondo, was open about its ETV focus, and he has now joined a private group that is publicizing ETV sightings (6). Information about these and similar earlier defense department programs has been published by former officials after more documents were recently de-classified (7,8). Documents that describe years of US legislative and executive branch interest in, and disinformation about, "UFOs" will be reviewed in *Lies*. But now, all that has changed: ETs and ETVs are no longer hidden behind the curtain of secrecy and disinformation that kept them out of view for seventy-five years. The facts are known and can no longer be ignored.

The ETV Evidence Changes Human Destiny

The ETV evidence changes the future of humanity. It shows us things that were seen by skilled people (pilots) and simultaneously recorded on electronic (radar and infra-red) sensors: things that are beyond human technical capacity. The American military name for these objects is "Unidentified Aerial Phenomena" (UAP). They used to be called "Unidentified Flying Objects" (UFOs). I call them

"Extraterrestrial Vehicles" (ETVs). because that is what they are. The rest of *Truth* reviews the history of our earlier encounters with ETVs, speculates about how they work, describes the ETs who pilot some of them, and explains what the ETs are doing with us and to us.

2. UFOs – History

The history of UFOs (now ETVs -- extraterrestrial vehicles) is so well-documented that the ETV reports recorded from the mid-1940s that have been collected in the many books cited in the *Notes* section of this book might take you a month to read. But *Truth, Lies and ETs* is a review and analysis, not an encyclopedia, and it might take you a day to read. This chapter starts by recounting some significant events in our UFO history. Then *Truth* moves on to describe the ETs, to discuss our interaction with ETs and to show how that interaction has changed the world.

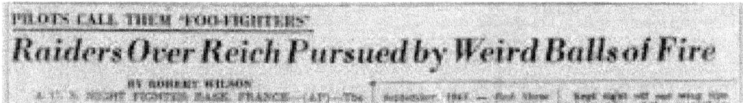

Figure 5. "Foo-Fighters" seen at the end of World War 2

Foo Fighters" and "Ghost Rockets" at the end of War II

ETV reconnaissance of Earth became routine in 1944-1945 when airmen on both sides of World War 2 reported that their airplanes were being followed at close range by luminous balls of orange or white light. The lights would move close to an airplane; the pilot's evasive maneuvers could not shake them off, and then the lights disappeared as suddenly as they had appeared. The lights were the same size as the aircraft or smaller. They came and went unpredictably. Figure 5 is a contemporary "Foo-fighter" newspaper headline. These luminous balls of light were never explained as either allied or axis secret weapons. They have not gone away: luminous balls of light associated with ETVs still interact with people and with aircraft. The ETV phenomenon is more complex now than it was in the foo-fighter era, but the invasive and persistent balls of light, the earliest consistent evidence of extraterrestrial presence in the twentieth century, are still part of the larger and more intrusive ET presence across the planet (9).

In 1945, soon after the German Army started launching V-2 rockets toward cities in England and liberated Western Europe during the closing months of World War 2, "ghost rockets" appeared over Finland, Sweden and Norway (10). The Ghost Rockets were first thought to be German V-2s launched from a German rocket test base that had been captured by the Russians towards the end of the war. Visual evidence contradicted the V-2 idea: many of the objects moved horizontally and much more slowly than either airplanes or rockets. Although observers had seen the objects dive into lakes, none were ever recovered. Radar tracked at least 200 ghost rockets, and the radar tracks matched the visual observations: objects that moved relatively slowly along non-ballistic trajectories. The radar and visual observations are consistent with what we now know – 75 years later – could have been ETVs, but the possibility of an extraterrestrial origin was not a major part of the conversation then. The official people studying ghost rockets did not discover their origin, but after the initial excitement was over, few people cared – World War II was over, much of the world was in ruins, and people everywhere were paying attention to things that were more important at the moment (Figure 6).

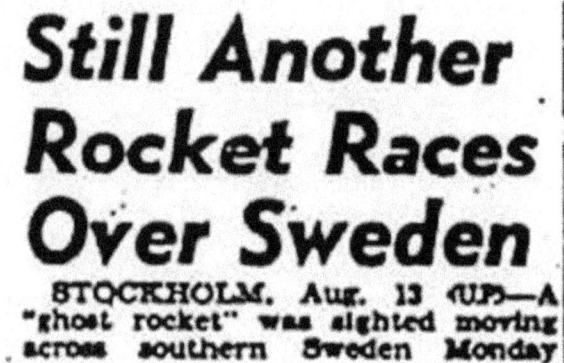

Figure 6. Contemporary newspaper clipping about "Ghost Rockets."

1947: Kenneth Arnold and "Flying Saucers"

On June 24, 1947, Kenneth Arnold was flying his private plane east across the Cascade Mountains of Washington State in the United States when a flash of light caught his eye. Then a second flash caught his eye. The flashes were reflected from a formation of nine bright objects traveling south between Mount Baker, far to his north, and Mount Rainier, closer on his left. They flew in echelon formation, with the lead object highest, as they passed over Mount Rainier and then disappeared over Mount Adams to the south. Knowing the distance between the mountain peaks, he estimated their speed at more than 1,200 miles per hour – faster than the few new jet planes of the era.

Arnold landed at Yakima, Washington and mentioned the sighting to people he knew at the airport, saying he had no idea what he had seen. His listeners suggested that he had seen guided missiles. Figure 7 shows his drawing of what he saw. Arnold then flew on to Pendleton, Oregon. Word of his sighting had preceded him, and when he landed he met a crowd of interested bystanders. Someone in the crowd had seen similar "mystery missiles" on the same day, and the opinion was again that Arnold had seen guided missiles. Arnold talked to a reporter from the *East Oregonian* newspaper. He remarked to the reporter that the objects' flight reminded him of a flat rock as it skipped across water. The term "Flying Saucer" was coined by an anonymous headline writer for the Associated Press, which circulated Arnold's story across the United States. The story triggered about twenty similar reports from the same day, almost all of them in the American Northwest. The Arnold story gave the world unidentified flying objects (UFOs then, ETVs now) and it made "flying saucer" part of our UFO vocabulary (11).

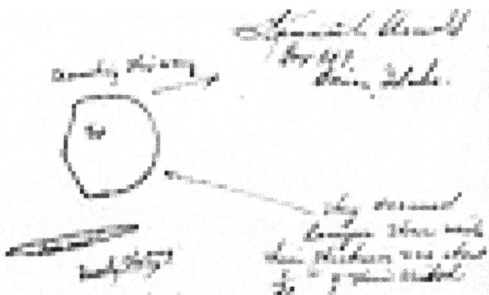

Figure 7. Kenneth Arnold's "Flying Saucer" seen on June 24, 1947

Newspaper accounts about the sightings in the Northwest gave little support to the idea of extraterrestrial origin. The papers acknowledged that the sightings were mysterious but other explanations predominated. Either the UFOs were misperceptions or hallucinations, or they were military technology—American or foreign. The Cold War between the Soviet Union and the West was underway in 1947 and no one, including the American military, knew what to make of these early ETVs. But a disinformation campaign was organized by the US government immediately after the many UFO sightings that followed Arnold's report, and it will be described later in *Lies*.

1950: McMinnville, Oregon

On the evening of May 11, 1950, Paul and Evelyn Trent saw, and Paul photographed, a UFO hovering over their farm in Oregon. Evelyn was outside feeding rabbits when she first saw it. She called Paul who was inside; he came out and saw a "brightly metallic, silver or aluminum covered" disc "with a touch of bronze" which made no noise and had no exhaust. They remembered that they had a loaded camera in the house. Paul ran back into the house, retrieved the camera, and was able to take two photos before the object flew off.

The Trents mentioned their sighting to friends but were not concerned about it. They thought it might have been a US secret weapon. When they finished developing the roll of film, Paul showed the photos to a few friends. The local newspaper heard about it and asked the couple for the negatives, which they provided. The newspaper published the photos on June 8, 1950, after which the story, as we would say today, "went viral" (only too appropriate, as this is being written during the 2020 covid-19 pandemic) and the "flying saucer" photographs became national news. Everyone, including neighbors and the editors of the local newspaper, vouched for the honesty and sincerity of the couple who saw and photographed the UFO (Figures 8 and 9). Sixteen years after the McMinnville photos were taken and reported, they were studied by a US government committee established to investigate UFO reports. The committee concluded:

> This is one of the few UFO reports in which all of the factors investigated, geometric, psychological, and physical appear to be consistent with the assertion that an extraordinary flying object, silvery, metallic, disc-shaped, tens of meters on diameter, and evidently artifactual, flew within sight of two witnesses (12, p 396 – 415).

The commentator did not "positively" rule out a fabrication, but wrote that many factors weigh against it. The commentator was Dr. William K. Hartmann, an astronomer working as a consultant for the Condon Committee, which was established by the US government in 1966 to evaluate the UFO evidence. The Condon Committee will be described in the *Lies* section. It was organized to investigate UFOs, it collected a lot of good evidence about UFOs, and then it lied about what it found.

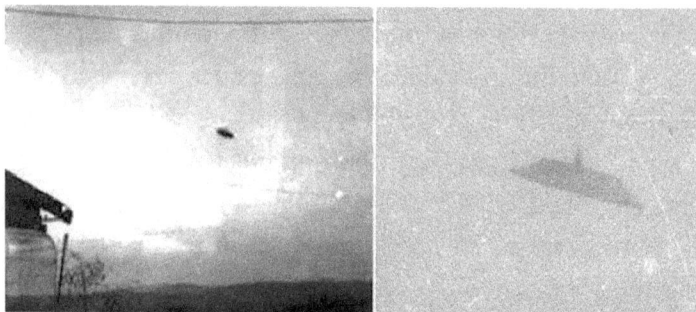

Figures 8 and 9. UFO photographed by Paul Trent near McMinnville, OR on May 11, 1950.

1952: The Washington, D. C. "Flap"

UFOs often showed up in groups that arrived in about the same place and at about the same time. Military slang – the "flap" – meaning "we have to do something about this, but we really don't know what to do" – has been used to describe those clustered UFO sightings ever since the US military started to react to them. The events described here became the first well-known "flap."

American UFO reports had peaked in July every year since the US Air Force began keeping records, which was immediately after Arnold's 1947 "flying saucer" sighting. July 1952 was a turning point for the public and for the government. On July 1, people in downtown Washington, DC looked up at something hovering northwest of the city. A physics professor saw the "dull, gray, smoky-colored" object and called the Air Force. The Washington, D. C. flap had begun (13).

UFO reports came in so fast in July that Air Force officers and staff worked fourteen hours a day, six days a week to deal with the American sightings, as well as keeping track of the many foreign reports that were also being received. Sightings were reported over a vital defense facility, the plutonium processing plant in Hanford, Washington State, and from many other places in the United States. And they continued to be seen over Washington, D. C. (Figure 11).

On the evening of Saturday, July 19, 1952, radars at Washington National Airport and nearby Andrews Air Force Base in Maryland tracked UFOs that flew slower than airplanes and then accelerated rapidly before coming back into radar and visual range. Airliner crews saw lights where the radars saw targets; jets were again scrambled to intercept, but as usual the interceptor pilots did not see, photograph, or shoot at anything .

The Air Force held a press conference about the Washington sightings at the end of July. The general in charge suggested that "temperature inversions" – an atmospheric phenomenon that sometimes affects radar reliability -- might have caused the sightings, and he was not questioned critically about it. He said the Air Force was continuing its investigations, but that the sightings did not constitute a threat to national security. After a month of press excitement, the Air Force explanation was taken at face value by most of the press. The Washington Flap was a turning-point in the US government's reaction to UFOs, and that change will be described in *Lies*.

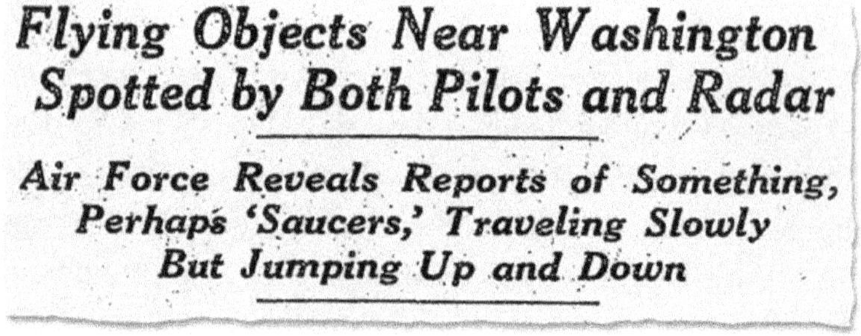
Flying Objects Near Washington Spotted by Both Pilots and Radar

Air Force Reveals Reports of Something, Perhaps 'Saucers,' Traveling Slowly But Jumping Up and Down

Figure 10. New York Times headline about the 1952 Washington, D. C. "Flap".

1957: The RB-47 Case

On July 17, 1957, a UFO stalked a US Air Force RB-47, an electronic reconnaissance bomber, for one and one-half hours over an 800-mile course from the Mississippi Gulf Coast to Oklahoma. The RB-47 tried to catch the UFO twice but it was outrun or outmaneuvered both times. The UFO was seen by the crew, detected by the RB-47's sensors, and tracked by a ground radar station that recorded the RB-47 and the UFO at the same time.

The UFO appeared first as an invisible radar return that was tracked by the RB-47's airborne radar receivers. The invisible return appeared to follow and then circle the RB-47 as it flew over Gulfport, Mississippi. As the aircraft flew further inland, the radar return suddenly appeared to the crew as a "very intense white light with [a] light blue tint" that flashed across the airplane's flight path and took up a position to the right side of the aircraft before it blinked out. The radar return moved with the light, and even after the light blinked out, the radar return never disappeared from the RB-47's radar receivers. The "huge light" then reappeared at the same bearing as the radar return and below the RB-47.

The RB-47 captain asked for and got permission from air traffic control to deviate from his flight plan and chase the radar-tracked light. He went to full power and dived, but as the RB-47 gained, the object suddenly stopped in mid-air below him and the RB-47 overshot it. He turned back towards the object, but when the RB-47 got within five nautical miles, the UFO dropped lower and both the RB-47 and the ground radar station lost contact with it. Running low on fuel, the captain radioed the ground controllers that he had to set a course for his home airport (in Topeka, Kansas), at which point the UFO took up station behind the RB-47 and was seen on the airplane's radar receivers until the RB-47 had passed Oklahoma City on its way home, when the radar signal finally faded out. Figure 12 traces the path of the RB-47 during its close encounter with the UFO.

This case has an official pedigree: it was assembled from flight plans, mission reports, and communications transcripts that were retained as military records. There is no doubt about witness reliability because the six aircraft crew members plus the ground-based radar tracking and flight control crew were all either military personnel or government employees whose careers depended, at the very least, on reliability and credibility. Separate follow-up phone interviews with the six flight-crew members were also consistent with the radar tracks and the airborne communications.

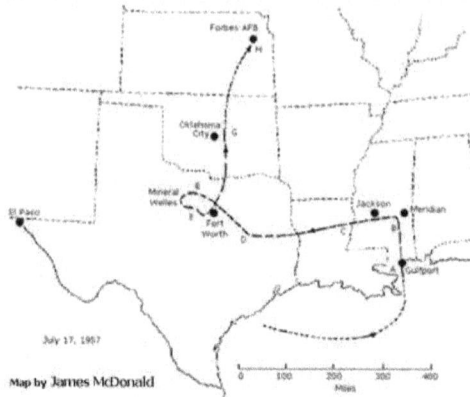

Map by James McDonald

**Figure 11. Where a UFO outmaneuvered
a USAF RB-47 in 1957**

Ten years after the Kenneth Arnold UFO sighting, a fast, maneuverable object, seen visually and reported on (radar, stalked and outmaneuvered a frontline US military jet aircraft. The malfeasance or misperception explanation cannot explain this case. The alternative explanations are that the visual and radar observations were caused by a previously unknown atmospheric-meteorological event, a secret military craft not known to the airmen and the radar operators, or an extraterrestrial vehicle. The atmospheric-meteorological explanation is weakened by the fact that no one knows how the atmosphere and the weather could produce such an effect. The secret military explanation is weakened by two arguments: first, it wasn't "secret" if it was playing tag with a frontline military aircraft where it could both be seen and tracked on radar; and second, no terrestrial machine more than a half-century after this event can even approximate the UFOs performance. That leaves the third explanation: the object was an extraterrestrial vehicle – an ETV (14).

1965: The Incident at Exeter

Norman Muscarello was hitchhiking home to Exeter, New Hampshire along a deserted road at about 2:00 a.m. on the morning of September 3, 1965. He watched a giant, egg-shaped, luminous thing with pulsating red lights around its circumference rise from behind the trees bordering a nearby field. It flew over him and away. He was terrified. He got a ride to Exeter and ran into the police station to report what he saw. An officer drove him back to the site in a police car, and they explored the field and the bordering trees. At first they saw nothing, but as they returned to the car the object reappeared. The officer shouted, "I see it!" into his open mike just as another officer pulled up in a second car. Both officers and the original witness watched the object fly off at treetop level toward the east.

A few minutes later, a hysterical man called the police from a phone booth east of Exeter to report that a huge, low, luminous object had just flown over his car. Earlier that evening, a cruising officer had found a woman sitting terrified in her car by the side of the road. She described seeing the same thing, but the officer, assuming she was hallucinating, had ignored her report.

Officers from nearby Pease Air Force Base, then home to B-47 and B-52 squadrons, made discreet inquiries about UFO sightings in the local area. They committed nothing compromising to paper, but they did not dismiss the evidence. Curious people congregated in places where UFOs had been spotted, hoping to see one for themselves. All of the local witnesses knew the difference between the B-47s and B-52s, which they saw and heard regularly, and the UFOs that they were now seeing.

The journalist John L. Fuller interviewed Muscarello and many other witnesses who had seen UFOs in and around Exeter over the previous months. Fuller visited one of the well-known UFO observation spots. With other witnesses he watched a jet fighter, identified by its noise and its running lights, chase a luminous UFO at an apparent altitude of 6–8,000 feet. Local witnesses reported many fighter-UFO chases over the previous few months. Fuller wrote about all of this in his book, *Incident at Exeter,* published in 1966. His summary of the UFO phenomenon as he observed it in the latter months of 1965 is complete and unambiguous:

- Dozens of intelligent, reliable people reported UFO sightings, many reluctant because of the fear of ridicule

- Most of the sightings were similar in description, and the police and military were reporting the same type of phenomenon as the ordinary layman.

- The reports of electromagnetic effects on lights, ignition, radios, and television indicated a similar conclusion.

- Photographs checked by experts, with full character investigation of photographer, added further evidence.

- The verified cases of genuine shock and hysteria indicted further that low-level near-landing reports were valid.

- Radar reports and scrambling jets chasing the objects indicated that the Air Force was not only cognizant of the objects but appeared to be impotent when it came doing anything about them.

- The most logical but still unprovable explanation is that the unidentified flying objects are interplanetary spacecraft under intelligent control (15).

1973: The Coyne Helicopter Incident

An Army Reserve helicopter almost collided with a UFO over central Ohio on the night of October 18, 1973. The four-man crew, commanded by Reserve Lieutenant Lawrence Coyne, was flying their UH-1 "Huey" medevac helicopter 125 miles southwest from Cleveland to Columbus, Ohio, to complete routine medical check-ups. As they returned to Cleveland that night the sky was clear and the helicopter was flying at 1,700 feet above rolling terrain.

At about 11:00 p.m. over Mansfield, Ohio, one of the crew saw a red light on the southeast horizon and told Coyne, who said "keep an eye on it." A few seconds later the crew member said the light was moving toward them on a collision course. Coyne put the UH-1 into a 500-feet-per-minute descent. He called the Mansfield tower to check on nearby traffic but got no response (a follow-up found no other civil or military airplanes in the area). The red light kept approaching, and Coyne kept the helicopter in a descent until they had just about reached the ground—about 650 feet up. As the object reached the UH-1, it stopped and hovered above and in front of the helicopter. A cigar-shaped, domed object almost filled the front windshield. It had a red light at the bow, a white light at the stern, and a movable beam that swung from the bottom of the object over the nose of the helicopter and into the cockpit, filling it with green light. Figure 12 shows what Coyne and his crew saw.

The object stayed over the helicopter for an estimated ten seconds and then sped off to the west and eventually out of sight to the north in the direction of Lake Erie. As the object left, the helicopter's magnetic compass swung wildly and the

UH-1 climbed to an altitude of 3,500 feet even though Coyne was still holding the control lever for descent.

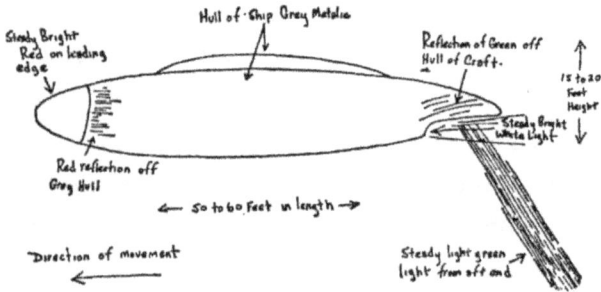

Figure 12. Sketch of the UFO that intercepted a US Army helicopter over Ohio in October 1973.

A family of four was driving on a nearby road below. They saw a red and a green light descend rapidly toward their car. They stopped and got out of the car to look at it. They first heard and saw the helicopter approaching from the south. Then they saw the red-and-green lighted object, which they said looked like a blimp and was as big as a school bus, hovering over the helicopter. The green light suddenly got brighter and lit up the helicopter as well as the family on the ground beneath. Now thoroughly frightened, the family scrambled back into their car and drove off.

Was this a misperception or a hoax? Military officers and flight crew do not report hoaxes. The independent ground witnesses and the flight report filed by Lieutenant Coyne agree about the details of the encounter. All four helicopter witnesses agree both about what they saw and about their maneuvers before and after the encounter. And there was physical evidence: the helicopter's magnetic compass failed immediately after the encounter and had to be replaced.

Was the UFO a secret military or civil aircraft? Only if the United States or a foreign power was testing a secret aircraft in the middle of the night over Ohio and while doing so, dangerously interfering with a routine military helicopter flight. It was seen at close range and was not recognized by any of the witnesses as a military or civil aircraft. Its performance was extraordinary. Years later no existing aircraft can do what that ETV did (16).

1982-1985: The Taconic parkway sightings

We prefer to be awake during the day, and as far as things that happen at night are concerned, most of us would prefer to ignore them. Night-shift workers put in hours while the rest of us are asleep. Just the same, if something happens after dark, it is not as much a part of the everyday world as if it had happened in broad daylight. If what is about to be described had happened in broad daylight, no one today would doubt the existence of ETVs.

Starting in 1982, low, big, and slow UFOs were seen at night (by one of my McGill University students, among many others) in the northern suburbs of New York City close to the Taconic State Parkway. UFO investigators were at work from the beginning. Philip Imbrogno, a high school science teacher, organized a team of local investigators associated with a private UFO research organization called the Center for UFO Studies (CUFOS). The team tracked down leads, recorded witness observations, and cross-checked witnesses' accounts against other witnesses, police records and newspaper reports.

Here is a witness report from March 17, 1983:

It was about twenty minutes to nine [p.m.], and I was driving home from a church meeting in Brewster [New York]. As we approached the house I saw a large, triangular object hovering over my yard about fifty feet from my house. It seemed to be not much higher than my roof. We pulled into the driveway, and we all jumped out and ran into the backyard. The object was no longer there. I took the children into the house to get them ready for bed, but I felt a strong urge to go back outside. As soon as I left the house, I saw the object hovering over I-84, just one hundred yards or so away and twenty feet or so above a truck that was passing underneath it. I ran in and got my children and my father, and we started to watch it. It now seemed to be just above a truck that had pulled over to the side of the highway. I was amazed to see how low the object was. The traffic was stopped, and people were out of their cars looking up at it. You could see people on the bridge pointing at it. I remember saying to myself "I wish I could get a better look at it." And as I was thinking that it made a 360-degree turn, as if rotating on a wheel, stopped, and started to float in my direction. It continued to approach me, and I just stood there transfixed. It stopped forty feet from me and was hovering twenty feet above a telephone pole in front of my house. It was a very large, V-shaped object, a very massive size. I watched it from the time it left the I-84 until the time it hovered, for approximately three minutes, and at that point all the lights seemed to intensify...I watched it alone with my dad for approximately another two minutes, when we started to walk underneath it. It seemed to be about the width of a football field and was a dark, very gray metal. It was so close you could hit it with a baseball.

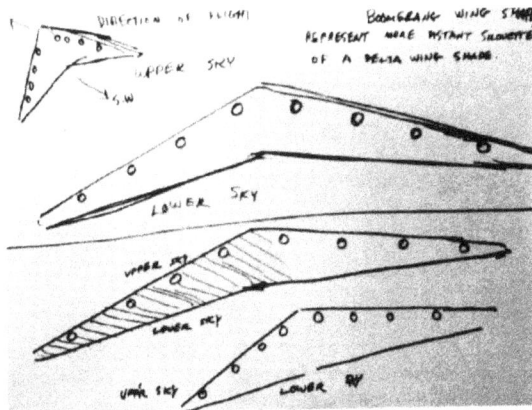

Figure 13. Low, big, and slow UFOs seen by hundreds of people over the Taconic State Parkway in New York state on March 24, 1983.

Imbrogno and his team gathered hundreds of additional witness reports that established a descriptive chronology from the end of 1982 to the middle of 1986, complete with photographs, of thirty-seven separate, well-corroborated, low-level nighttime appearances of giant, brightly lit, boomerang-shaped or triangular UFOs in the Taconic State Parkway region and nearby parts of New York and Connecticut (Figure 14). Imbrogno and his coauthors documented later sightings from the same area in the second edition of their book. Their follow-up covers cases from 1986 through 1995, many of them similar to the UFOs described their first edition, and the work is a tribute to their continuing dedication and perseverance.

The US Federal Aviation Administration explained that the Taconic Parkway UFOs were "stunt pilots flying at night (17)."

1989: Belgium

Low, big, and slow UFOs were seen at night over Eupen, a small Belgian town near the German border, towards the end of 1989. The reports described things that looked like those seen along the Taconic State Parkway. They were seen by police, civilians, off-duty Air Force officers, and "several witnesses [who] had high-ranking functions and preferred not to reveal their names to the media" (16, p. 34.) But plenty of witnesses with medium- to low-ranking functions were willing to be named.

The government reacted quickly. They alerted the Ministry of Defence, which assigned the investigation to Wilfried De Brouwer, chief of operations of the Air Staff.. The Belgian Air Force scrambled fighter jets on two occasions when police reported low-level sightings that were confirmed by radar. The first time the jets found nothing. The second time one of the jets recorded a camera sequence of its radar lock-on of the UFO. Ground and airborne radar reconstruction of that jet's weaponless dogfight with the UFO showed a large object leading the jets on a chase before speeding out of sight and out of radar range. The objects seen in the Belgian sightings and the Taconic State Parkway sightings were similar (Figure 15). Neither set of observations could be explained as machines made by humans. They were machines made by someone else. They were ETVs (18).

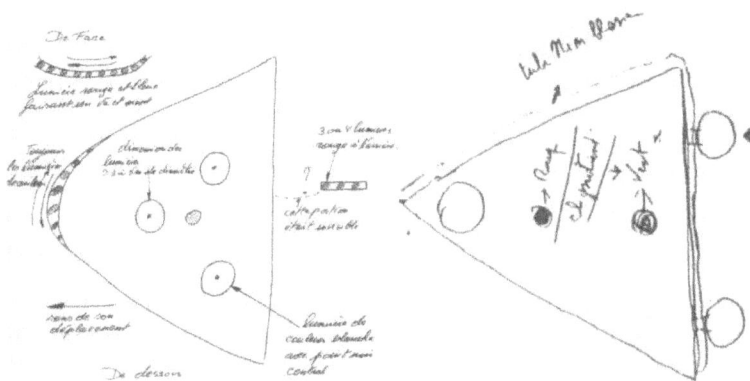

Figure 14: Witness drawings of ETVs seen near Eupen, Belgium in 1989.

1996: Yukon Territory

An enormous object—much bigger than those reported near the Taconic State Parkway or in Belgium—showed up over the Yukon Territory in Canada on the clear night of December 11, 1996. This single huge object moved slowly enough that successive reports from witnesses at different locations established a record of its path, its size, and its speed.

The Belgian UFOs were seen in a small corner of a small country. The Yukon UFO was seen over a large part of a large country; for about 205 miles along the Klondike Highway. The UFO was seen, sometimes directly overhead and sometimes at a distance but still low in the sky, by at least thirty- one people -- some along the highway near Fox Lake, others further north in the villages of Carmacks, Pelly Crossing, and Mayo (Figure 15).

Martin Jasek is a civil engineer and a private UFO

investigator. Jasek went to the Yukon Territory and talked to witnesses who told him about other witnesses. He collected interview reports, took witnesses to their observation sites, asked for position and dimension estimates, and asked them to sketch what they had seen. Jasek analyzed the individual witness accounts, correlated witnesses' direction and dimension estimates with topographical map features, and constructed an estimate of the size and appearance of the UFO by triangulation from the many observers' estimates. His computer-aided rendering of the reported UFO is shown in figure 16. The average triangulated diameter of this round UFO was nearly *one mile*, and its estimated height from top to bottom was one-third of a mile. Witnesses who saw it overhead from the Klondike Highway estimated that it was sometimes no higher than 250 feet off the ground (19).

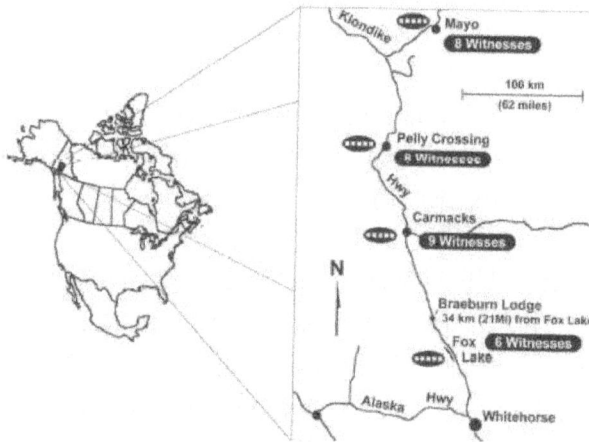

Figure 15. Location of the witnesses who saw the 1996 Yukon Territory UFO.

Figure 16. UFO seen in 1996 over the Yukon Territory, Canada

2008: Stephenville, Texas

A giant UFO flew low over Stephenville, Texas in January 2008. Eyewitnesses included a constable, a chief of police, a former Federal Aviation Administration (FAA) air traffic controller and a private pilot. Radar experts, using air traffic control radar plots from several nearby civil airports, confirmed that a low-flying target, without a civil or military aviation transponder, hovered and then flew at high speed across several Texas counties, and that military jets—some fighter aircraft and some tracking aircraft—crisscrossed the Stephenville area shortly after the UFO was seen there. The US Air Force admitted that some military jets were in the air but otherwise said nothing about the sighting. Neither did the FAA, except to provide the air traffic control radar plots on request, as required by law. The reports and analysis describe an object that was between 500 and 1,000 feet long, observed at an altitude of between 5,000 and 17,000 feet, and that sometimes flew very slowly (~50 mph) or very fast (~1,900 mph). Investigators Glen Schulze and Robert Powell wrote "It was not any known aircraft" (Figure 17), (20). It was an ETV.

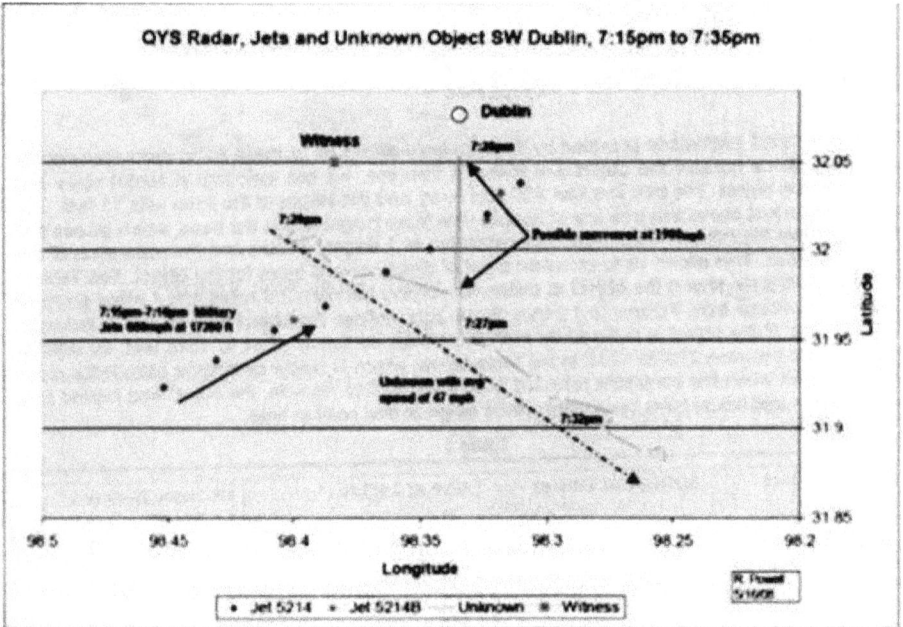

Figure 17. Radar plot of ETV seen, intercepted and recorded over Stevenville, TX in January 2008

The End of Doubt

Human experience, which is constantly contradicting theory, is the greatest of truth. Samuel Johnson's observation headlines the *Truth* section. It is the foundation of truth in the world and it is the foundation of the truth told in this book. Observations come first; theories to explain the observations come second. The corollary of Johnson's observation is that the absence of a theory does not diminish or cancel the reality of human experience.

The ETV evidence consists of human visual observations corroborated by instrumental records. Accepting the ETV evidence was at first controversial, because the prevailing "scientific" explanation was that observers misperceived natural phenomena and then psychologically altered their memories to fulfill a need to claim esoteric knowledge. These first two chapters should have disabused you of that nonsense. The visual observations and instrumental records are reports of ETVs.

The information reported throughout *Truth, Lies and ETs* is based on human experience supported by instrumental evidence, and by psychological research that supports the human evidence. If you are a skeptic, your job is to find alternative explanations for the evidence. If you ignore the evidence then you ignore what human observation tells you about the world, and your opinion deserves to be ignored in turn.

Independent and consistent observations prove that ETVs are real. Observations have been made by civilians, police officers, military officers and pilots. Visual sightings have been correlated with ground radar plots of the objects. Fighter pilots have chased objects that were observed visually and plotted on ground and airborne radar. The objects were often larger than aircraft and they were maneuverable.

No one who has read and understood the evidence should now doubt that we have seen and tracked machines in our skies: machines that we do not know how to make. It is more accurate to say that we *have been shown* machines that we do not know how to make. The ETVs reported over Europe during the Second World War, over The Cascade Mountains in Washington State in 1947, over Washington, DC in 1952, over the Gulf Coast of Mississippi in 1957, over Exeter, New Hampshire in 1965, over central Ohio in 1973, over the Taconic State Parkway region of New York in 1982, over Belgium in 1989, over the Yukon in 1996, over Stephenville, Texas in 2008 and over the Pacific and Atlantic Oceans from 2004 to the present were not hiding. They were on display for us.

3. How do ETVs work?

I'm a research psychologist and engineering physics is not part of my professional skill set. My professional answer to: how do ETVs work? has to be: I don't know. But here are some suggestions..

Scientists have been experimenting with gravity, a phenomenon of nature that is related to how ETVs work, since long before ETVs were in the news. Their experiments may offer us some clues about ETV performance. There are also more recent experiments that relate to the problem of how ETVs travel long distances at high speed, then stop, hover and start apparently instantaneously – performance that is beyond what we, at least publicly, know how to do. But there are some publicly accessible patents, one of which is owned by the United States government, that relate to ETV propulsion.

Some of the scientific research, the "ETV patent" and some private patents give us just a hint about how ETVs might work.

Thomas Townsend Brown' Gravitator

Thomas Townsend Brown (1905-1985) was an American inventor who thought that gravity could be modified by both the sign (positive or negative) and the strength of an electric charge that was held within and on an object. He predicted that an electrically positive charge at the surface of an object would move the object against the restraining effect of gravity and inertia. (Inertia is the resistance of an object, based on its mass, to any change in its condition of either rest or uniform motion).

According to Brown, if you put an electrically charged object on a scale with the positively charged side up and the negatively charged side down, the object will pull itself a little bit off the scale and so weigh less; if you put the object on the scale with the positively charged side down and the negatively charged side up, it will push itself onto the scale and so weigh more. This experimental result is not explained by currently accepted theories of physics, but alternative explanations that stay within accepted principles have been proposed.

The device that he first patented in the United Kingdom (21) is called the Gravitator. When the Gravitator (Figure 18) is positively charged to the right, the entire object moves to the right inside the oil bath that surrounds it. If the charge is reversed so the positive charge is to the left, the entire object moves to the left. Thus, according to Brown, gravity and inertia, the forces that regulate our motion on the surface of the earth, can be modified by changing the distribution of the electric charge on an object. On any object.

Modern critics argue that Brown's results are caused because the electric charges on the Gravitator generate opposite electric charges on nearby air (or oil) particles, which then push away the Gravitator and make it move. Brown' research has been replicated in a vacuum (22), but modern researchers still find reasons to question whether he demonstrated a straightforward relationship between gravity and the distribution of electric charge. Whatever the explanation, Brown's research and his patent generated a flood of research about how to modify gravity, and that research continues to this day.

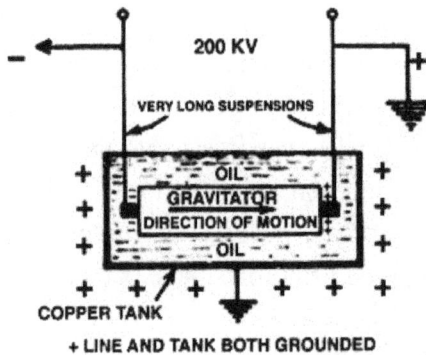

Figure 18. Thomas Townsend Brown's Gravitator

Hideo Hyasaka and Sakae Takeuchi's Spinning Disc

Hideo Hyasaka and Sakae Takeuchi put a magnetic disc into a vacuum jar and weighed it. Then they spun the disc very rapidly, at velocities of between 3000 and 13000 rpm. They weighed the disc as it rotated and found that it when it was spinning clockwise (viewed from above) it weighed less, and that the weight loss was directly proportional to the speed of rotation. Spinning a magnetized object generates an electric charge and an electric current. Hyasaka and Takeuchi showed that a change in the mass (the weight) of an object was caused by a moving magnetic charge that generated an electric current. They protected themselves against the criticism of Brown's work (that the movement of his device resulted from the repulsion of electrically charged air particles) by carrying out their experiment in an airless vacuum jar. Hayasaka and Takeuchi's disc, spinning in a vacuum, lost weight, demonstrating a relationship between an electric charge in motion and gravity (Figure 19), (23).

This controversial experiment has been replicated successfully with various modifications, but its result: the loss of weight by a spinning electrically charged disc, is still not explained by any widely accepted theory (24, 25). There are other experiments showing that both the weight of an object and its inertia (its resistance to being accelerated) can be modified by the presence of electric currents. Each research group has its own slightly different theoretical explanation for why this happens and there is as of yet no accepted theory that explains how it works.

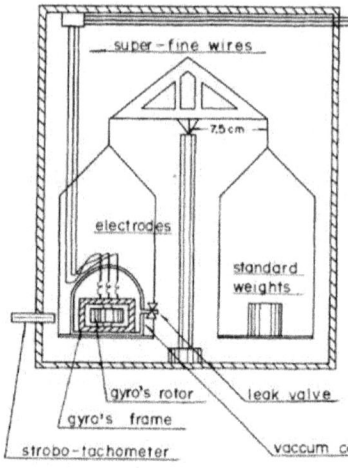

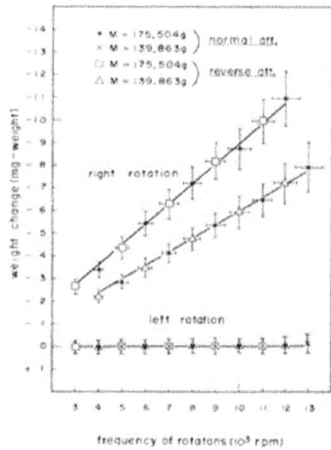

Figure 19. The Spinning disk that gets lighter as it spins faster. (Physical Review Letters, 63(25), 18 December 1989)

Hector Serrano's Propulsion Device

Theory or not, engineers have gone to work on ETV technology and some of their work has been recognized by the U.S. Patent Office. .On December 10, 2002, Hector L. Serrano was awarded US patent 6,492,784 B1 for a "Propulsion device and method for employing electric fields for producing thrust." (Figure 20). Does that sound familiar? It should. It is what Thomas Townsend Brown's Gravitator patent claimed to do. Serrano used a different and more sophisticated method. His patented device also has something in common with the spinning disc tested by Hyasaka and Takeuchi that was just described. This patent is for a *thing* –

not for an experiment or an idea, but for something that works. Serrano's "propulsion device" exists both as a drawing (Figure 21) and as working device (Figure 22) When Serrano's propulsion device – a more sophisticated version of Thomas Townsend Brown's Gravitator, is charged, its weight is reduced, just like Brown's Gravitator and Hyasaka and Takeuchi's spinning disc. The Gravitator, the Spinning Disc and the Propulsion Device all suggest that gravity and inertia can be overcome by harnessing electromagnetism. Critical experiments suggest the effects are small in a vacuum, and there is debate about whether the apparent effects on weight depend, not of a change in gravity, but of the interaction of electric charges with the surrounding air.

These were all "desktop" examples. However it works, a desktop-size device will probably not get you to the stars. But a larger version might – and one was forthcoming in 2018.

US006492784B1

(12) **United States Patent**
Serrano

(10) Patent No.: **US 6,492,784 B1**
(45) Date of Patent: Dec. 10, 2002

(54) **PROPULSION DEVICE AND METHOD EMPLOYING ELECTRIC FIELDS FOR PRODUCING THRUST**

OTHER PUBLICATIONS

David, Leonard, *Warning. Mind Fields, Nasa Looks Into Breakthrough Physics*, Final Frontier, Jan./Feb. 1997, pp.

Figure 20: Hector Serrano's Propulsion Device Patent

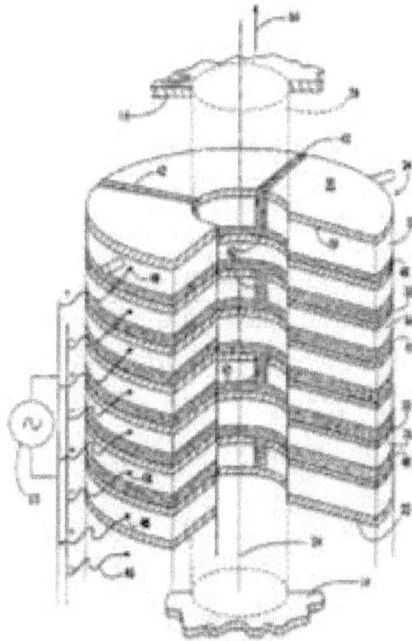

Figure 21. Hector Serrano's propulsion device

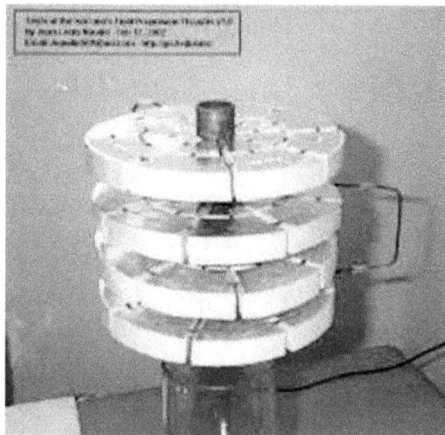

Figure 22. Working model of the Serrano device

Salvatore Pias' ETV patent

Could the effect of gravity on mass and inertia be weakened so much that we could leave the earth unrestrained by gravity? The U.S. Navy thinks so and has a patent that claims it can be done. US Patent 10,144,532 B2, for a "Craft using an inertial mass reduction device," was issued to Salvatore Cezar Pais on December 4, 2018 (Figure 23). Pais is a US Navy civilian scientist and his patent development rights belong to the US Navy. Pais cites the work of Hyasaka and Takeuchi as a precedent. His patent then lays out a different method for manipulating the electromagnetic field around an object (in Pais' patent, a "craft"). His claim asserts that a craft built to his patent's specifications would maneuver easily though the local field of low gravity produced by his patented technique. The craft, according to the Pais patent application, would achieve very high velocities and accelerations without endangering the people inside because they too would be protected by the reduced gravity, and therefore they would neither feel nor be injured by the acceleration. Just like in an ETV.

US010144532B2

(12) **United States Patent**
Pais

(10) **Patent No.:** **US 10,144,532 B2**
(45) **Date of Patent:** **Dec. 4, 2018**

(54) **CRAFT USING AN INERTIAL MASS REDUCTION DEVICE**

(71) Applicant: **Salvatore Cezar Pais**, Leonardtown, MD (US)

ference on Future Energy, Oct. 9-10, 2009, Washington, DC, US.
Pais, Salvatore, Conditional Possibility of Spacecraft Propulsion at Superluminal Speeds, Intl. J. Space Science and Engineering, 2015, vol. 3, No. 1, Inderscience Enterprises Ltd.

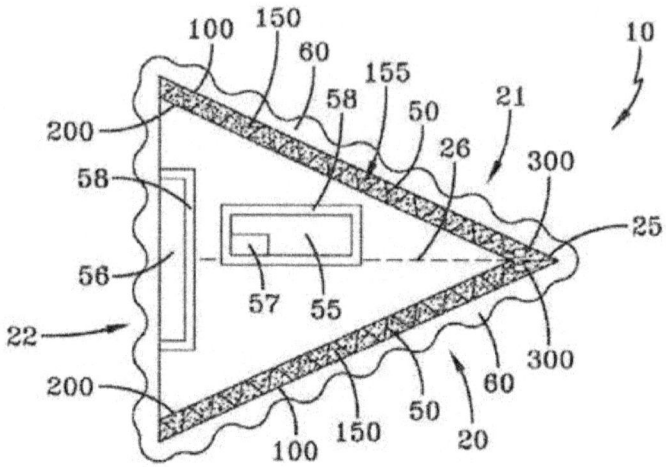

Figure 23. Salvatore Pias' patented triangular ETV.

A Note About Patents

The United States Patent and Trademark Office says that

> A patent cannot be obtained upon a mere idea or
> suggestion. The patent is granted upon the new
> machine, manufacture, etc. as has been said, and
> not upon the idea or suggestion of the new
> machine. A complete description of the actual
> machine or other subject matter for which a
> patent is sought is required (26).

This means that someone has proved to the Patent and

Trademark Office that the machines pictured in Figures 22, 23, 24 and 25 are complete – including the "Craft using an inertial mass reduction device" of Salvatore Pais – the anti-gravity ETV. It does not mean that a full-sized version has been built, because a patent can be issued based on a model. Canadian Patent 1275202, "Device for pushing broken ice ahead of a ship or the like," was issued to me in 1990, based on a model that I had developed while studying marine navigation in the arctic. The patent records show that we may know how to build an anti-gravity ETV; but not necessarily that we have built a full-sized one.

Is there a working version of Pais' patented craft? I don't know. We are at the edge of what governments will tell us. The military and economic advantages of controlling gravity are immense, and engineering progress towards antigravity propulsion is supported by governments and is mostly hidden by secrecy. But that secrecy was compromised a few years ago, and according to one observer, "reverse engineering" of ETVs has already happened.

The Alien Reproduction Vehicle

The late Mark McCandlish was a professional aerospace illustrator who routinely depicted design concepts for military aircraft propelled by rockets and jet engines. One day a technically trained and highly placed friend of his walked into an air show and entered, by accident, a screened-off security area. Three prototypes of advanced propulsion vehicles were on display – hovering silently and motionlessly above the floor. His friend remembered in detail what he had seen and described one of the vehicles to McCandlish, who then drew a version of what the friend had seen. The sketch, known as the Alien Reproduction Vehicle (ARV), (27) has been widely reproduced and is shown in Figure 24.

The Alien Reproduction vehicle may work on principles similar to those reported in the Hayasaka and Takeuchi experiment and the patented Salvatore Cezar Pais "Craft." It generates a rapidly fluctuating electromagnetic field that surrounds and fills the ARV and reduces is mass. The anti-gravity field is powerful enough to lift and propel a vehicle that looks something like a "sports model" ETVs (figure 27): but more about that later.

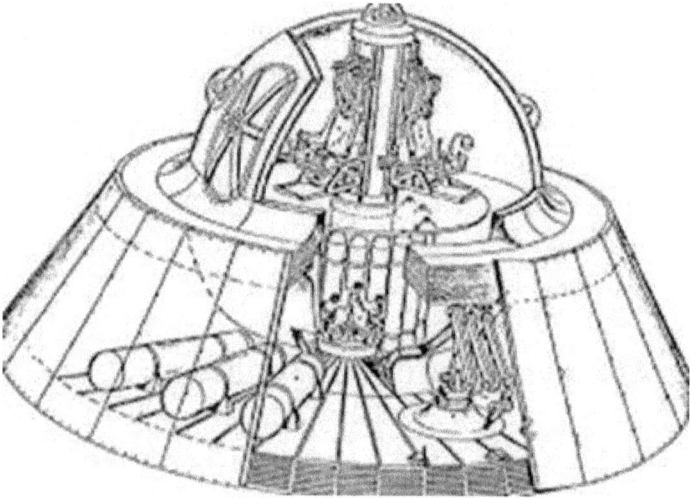

Figure 24. The Alien Reproduction Vehicle.

A Summary of how ETVs Might Work

This brief account of how ETVs might work reports one experiment and three patents. It ends with the drawing of an anti-gravity craft that was seen by a technically trained person as the result of a security lapse and then re-created from his description by a technical illustrator. Critics say that some of what is considered to be evidence for "antigravity" effects may just be an interaction between the electrically charged apparatus and electrically charged particles in the air – and not likely to get us to the stars. A skeptic could argue that neither the experiments, nor the patents, nor the security breach about the Alien Reproduction Vehicle prove "beyond reasonable doubt" (the standard for conviction in a criminal case) that we know how to make a machine that does what ETVs have been observed to do, and I agree. But these experiments and patented devices do show us that we *may* be able to exploit an interaction between electromagnetism and gravity that theorists do not completely understand. Engineers may not understand that interaction either, but they may have already built machines that work like ETVs, even though there is no complete and widely accepted theory about how they work. A provisional answer to the question: "How do ETVs work?" is that ETVs might have some means of modifying gravity and inertia that allows them to accelerate and decelerate rapidly without damaging either the ETV or the living creatures inside, and that this modification also allows ETVs to travel extremely fast for very long distances.

The human experiences described in this chapter, contradicting as they do some long-held scientific assumptions, have been accepted by academic peer reviewers and by patent examiners. They are *part* of Samuel Johnson's "human experience" that is "constantly contradicting theory" and that is "the great test of truth." The story is incomplete –

it does not tell us *how* ETVs work, it tells us how they *might* work. The complete story will be our admission ticket to the universe on terms of technological equality with its other civilizations. So much is at stake in mastering ETV technology that the only people who know much about the whole story are the people at work in the labs and shops that are dedicated to making "our" gravity-defying ETVs. The rest of us must be content for now with the partial story outlined by the publicly reported experiments and patents, and the report of an inadvertent security lapse.

4. Extraterrestrials

Roswell, 1947

On June 24, 1947, Kenneth Arnold flew over the Cascade Mountains in Washington State, saw ETVs, and "flying saucers" became headline news. Two weeks later, on July 8, the Roswell, New Mexico *Roswell Daily Record* reported that "RAAF [Roswell Army Air Force] Captures Flying Saucer on Ranch in Roswell Region" (Figure 49). The Roswell "capture," announced two weeks after the world first heard about flying saucers, marks the public start of the US government's confrontation with the reality of extraterrestrials. The facts of the Roswell incident have been assembled by persistent investigators who worked for years after the event to overcome the US government's effort to hide the Roswell crash. The disinformation part of the Roswell story will be described later in the "Lies" section of this book. But in New Mexico in 1947, we meet ETs at close range for the first time.

On Thursday, July 3, 1947, W. W. "Mack" Brazel, foreman of the Foster Ranch near Corona, New Mexico, was riding on the ranch when he discovered an area about 200 to 300 feet wide and three- quarters of a mile long, oriented northwest to southeast, and densely covered with irregular pieces of strange material. There was a shallow new gouge in the ground about 500 feet long and ten feet wide, starting at the northwest corner of the debris field and continuing to the southeast.

The debris included very thin, parchment-like sheets that were so strong they could not be bent or cut. There was also a foil-like material which, if crumpled, smoothed out again by itself. There were pieces of what appeared to be metal with some plastic properties. The metal was light, dull gray and non-reflective. It could not be cut or burned. Hammering did not deform it. There was also something that looked like nylon monofilament fishing line. There were small, flexible and strong I-beams covered with symbols. The material was so thick on the ground that Brazel's sheep would not cross the littered area.

Brazel drove into Corona and mentioned the debris to a brother-in-law, who told him about the "flying saucers" that had been seen the previous week (Brazel had neither phone nor radio at his house on the ranch). The brother-in-law suggested that he take the material to the county seat at Roswell. Brazel picked up some of the material and showed it to his nearest neighbors, who declined his offer to ride out and look at the rest of the debris. They also suggested that he take some of it to the county seat at Roswell.

On Sunday, July 6, Brazel drove the seventy-five miles from

the Foster Ranch to Roswell. He showed the material to Sheriff George A. Wilcox, who called the Roswell Army Air Force Base. Base commander Colonel William H. Blanchard, Base intelligence officer Major Jesse Marcel and Counter-intelligence corps Captain Sheridan Cavitt showed up at the sheriff's office shortly afterward. Blanchard notified 8th Air Force Headquarters in Fort Worth, Texas about the debris, and acting on instructions from them, sent some of the material by air to Fort Worth. Surmising correctly that the gouge and the debris field found by Brazel was caused by a forced touchdown that preceded a crash, the Air Force carried out an aerial reconnaissance of the area. They soon found the crash: a small disc-shaped craft, crumpled and broken open, and four extraterrestrials on the ground outside the craft. Three of them were dead and one was alive.

The damaged saucer, the debris at the debris field and crash site and the ETs were trucked to the Army Air Force Base in Roswell. The base was the home of the 509th Bomb Group; the atom-bomb carrying squadron that ended the second world war by destroying the Japanese cities of Hiroshima and Nagasaki. Security at the base, with its B-29s and atomic bombs, was already high, and security got even tighter as the debris and the ETs were trucked in and stored in empty hangars and in heavily-guarded makeshift locations on the base. Mortuary supplies were requisitioned from a local funeral parlor. Additional military personnel were requisitioned for guard duty at the debris storage locations. Local civilians, involved as contractors for one or another of the base services were, sometimes accidentally and sometimes as part of their routine duties, involved with the storage and preparation of the ETs for transport. Many of them saw the debris and several of them saw the ET bodies.

Instructions from Washington were to ship the material and the ET bodies to Wright-Patterson Air Force Base in Dayton, Ohio, the headquarters of the Air Force Technical Command. In the process of storing and moving the crash material preparatory to loading it on B-29s and other transport aircraft, more civilians and lower-ranking military personnel assigned to the medical, storage, shipping and security details also saw the debris and the bodies.

The "official" end to the Roswell incident was the publication, a few days later, of a cover story, provided to the local press and to the entire country by the press wire services, that what Brazel had really seen were the remains of a crashed weather balloon. The coverup, documented by ETV researchers, will be described under "Lies."

The Roswell ETs

The RAAF base hospital called the funeral home with which they had a contract and asked them to deliver "children's caskets." The hospital also asked for advice about embalming fluid. Another call was made from the base hospital to a local dairy to supply dry ice.

Miriam Bush was the administrative assistant to the base hospital director. The hospital was in an uproar and she was upset that she did not know why. The hospital administrator, Lt. Colonel Harold Warn, took her to the emergency room to see for herself. She saw bodies on gurneys in the middle of the room. She said "My god! They're children." Then she saw that they weren't children. Their heads were too large. And their eyes were too large. And one of them moved.

Other witnesses described the ETs as being 3½ to 4 feet tall, with long arms and frail bodies. Their skin color was described as ranging from beige to blue-grey, and the texture was described as "scaly" or "reptilian," reminding observers of iguanas or chameleons. There were four fingers (no thumbs) on each hand. The heads were oversized, the eyes were sunken and oddly spaced, the mouth just a slit and the nose and ears just small openings. In other words: the Roswell ETs were the prototype for every contemporary rendering of a small extraterrestrial (Figure 25), (28).

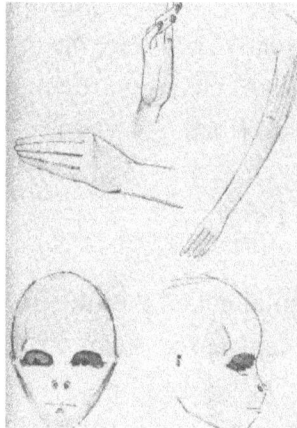

Figure 25. The Roswell ET, drawn from a description by a medical worker who saw the body.

Socorro, New Mexico: 1964

Police officer Lonnie Zamora was chasing a speeder in Socorro, New Mexico, at 4:15 in the afternoon of April 24, 1964. He "heard a roar and saw a flame in the sky to the southwest some distance away." Zamora knew there was a dynamite shack in that direction, so he abandoned the chase and went to investigate. As he drove closer to the shack he saw something he described as a large egg supported by slender legs sitting just off the ground. He saw two small creatures dressed in white coveralls standing near it. They appeared startled when they saw his car approach. His view was momentarily blocked by a hill and when he saw the object again, the creatures were no longer visible. He stopped his car and got out, intending to approach on foot, but the object emitted a roar and blue and orange flames pulsed from the bottom. Zamora ducked behind his patrol car. The roaring stopped and when he looked again the object was hovering a few feet off the ground. He saw it speed away, barely clearing the dynamite shack. Another officer pulled up and watched with Zamora as the object flew off (Figure 28). A third witness saw the object while it was in the air. The two officers went to the spot where the egg-shaped ETV had been parked. They found charred and singed grass, and indentations in the ground made by its legs.

Major Hector Quintanilla, who was in charge of the Air Force UFO study program called Project Blue Book, checked with aerospace companies, NASA, and other government agencies to learn whether they had anything operating in the vicinity; they all said no. There were no helicopter flights, aircraft or balloons in the area at the time. Quintanilla sent an Air Force consultant, astronomy professor J. Allen Hynek, to investigate. One of Hynek's scientific associates happened to know Zamora and gave the officer an outstanding character reference. There were physical traces. ETs were seen near the object. Hynek called it an extremely strong case and project Blue Book, the Air Force's ETV investigation organization, decided to label it an Unknown (but confidentially, following established policy). According to UFO researcher and scholar David Jacobs, "This is the only combination landing, trace and occupant case listed as unidentified in Blue Book files." The reliability and clarity of Zamora's ET observation and the testimony of other witnesses who also saw the ETV make this a well-documented observation of ETs next to a landed ETV (Figure 26), (29).

Figure 26. Rendering by M. Thompson of what Lonnie Zamora saw In the New Mexico desert in 1964

Ririe, Idaho, 1967

In 1968 a UFO investigation group called the National Investigations Committee on Aerial Phenomena (NICAP) asked a panel of aerospace medicine physicians, anthropologists, astronomers, biologists, clinical psychologists, research psychologists (including me), linguists, neurologists, philosophers, psychiatrists, and sociologists to evaluate six ETV "occupant" reports that had originated with local NICAP committees in the United States.

A maintenance worker saw occupants near a landed ETV in a municipal park at night. Other reporters were a dairy farmer spreading manure on a pasture, a factory worker who was hunting in the woods, an industrial draftsman who was chopping wood at an archery range, and a twenty-year-old man driving home from work. All of them reported seeing short humanoids, similar to the ones recovered in Roswell, sometimes in pairs and sometimes in groups. Most of the witnesses were terrified by their encounter, and the people who saw the witnesses soon afterward described them as "white as a sheet" or used similar terms.

One case is particularly detailed because the witnesses told their close encounter story to the police. Willie Begay and Guy Tossie were twenty-three-year-old Navajos who worked on a ranch near Ririe, Idaho. On November 2, 1967, they finished work and went to a local bar where they had a couple of beers, then got into their car to drive home. They had driven about a quarter-mile outside of town when a flash of white light startled and temporarily blinded them. They saw a small ETV (Figure 27) hovering about five feet off the ground in front of their car. It was about six feet in diameter and about three feet high and had a transparent bubble top through which they could see two occupants.

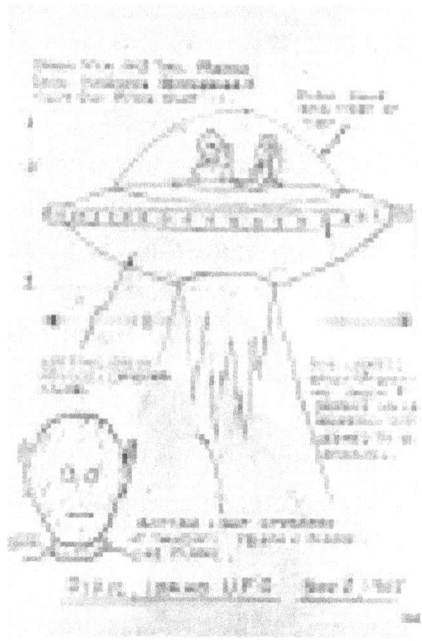

Figure 27. Occupants seen by Guy Tossie

Willie Begay, who was driving, said that even though he did not touch the brakes, steering wheel or accelerator, the car was stopped, driven off the road and then stopped again in an adjacent field about seventy-five feet from the road. The ETV top opened and one of the occupants floated down from the ETV, opened the driver's side door, and got in behind the steering wheel. Both men were terrified. While Wille Begay slid away from the occupant behind the wheel, Guy Tossie opened the passenger side door and ran to a nearby house, where he pounded on the door until the homeowner, Guy Hammond, let him in.

At the Hammond house, Tossie, frightened and incoherent, babbled something about a light driving their car off the road and that his friend was dead. Hammond and his son persuaded Tossie to accompany them back to the car, where they found Begay sitting in the car with his eyes closed with the car engine running and the lights on.

Hammond, meanwhile, returned to the bar in Ririe to see if the bartender could shed some light on the witnesses' state of mind. While he was talking to the bartender, the local constable and the county deputy sheriff came in for a sandwich. Then Begay and Tossie, who had been too afraid to drive home alone, returned to the bar for something to "settle their nerves." As soon as they saw the sheriff and constable, they rushed over to tell their story. The sheriff and constable contacted the Idaho State Police, who sent an officer to investigate. The officer found car tracks in the field but did not find marks or above-background radiation on the car.

The investigator located and interviewed another witness who had seen an orange light, low in the sky, flying in a zigzag pattern in the same area at about the same time that evening (29). Their report shows ETs in a small two-seater machine like the ETV "sports roadster;" the cartoonist's "standard model" that hovers over unsuspecting people while they are gardening or out for a walk (30, 31).

North Bergen, New Jersey, 1974

Budd Hopkins, who was an American painter and sculptor, lived in the Chelsea neighborhood of Manhattan in New York City. One November evening in 1974, Hopkins crossed the street to George O'Barski's liquor store to buy a bottle of wine for dinner. He found O'Barski, whom he knew well, pacing up and down behind the counter of his small establishment, muttering about life's indignities: "a man can be driving home, minding his own business, and something can come down out of the sky and scare you half to death." O'Barski's story was interrupted by other customers and by Hopkins' waiting dinner. Hopkins, his wife, and a friend had seen an

ETV on Cape Cod in 1964, so Hopkins was interested enough to go back after dinner and hear the rest of O'Barski's story.

On an unseasonably warm night in January 1974, O'Barski, a teetotaler, closed his store at midnight, restocked the shelves, did some bookkeeping, walked his guard dog, drove through the Lincoln Tunnel to an all-night diner in Fort Lee, New Jersey and then back towards his home in North Bergen. His driver's side window was rolled partway down. As he drove south past the east side of North Hudson Park in North Bergen, just a few hundred yards from the Hudson River, his radio began to pick up static. He heard a humming or droning sound from outside and saw a brightly lit object fly low from the left ahead of his car and towards the park. O'Barski slowed as the road curved, bringing into view a thirty-foot-long craft with a row of tall, narrow windows that had landed in the park on a field to his right.

He watched as a panel opened between two of the windows, a ladder descended, and nine or ten small figures climbed down the ladder to the ground. They were between three- and-a-half and four feet tall. Each wore an identical helmeted, one-piece light-colored garment and each carried a little bag and a spoon-like tool with which they dug into the soil around the craft and put it into bags that they were carrying. The creatures paid no attention as O'Barski watched them for an estimated four minutes. He saw the craft rise and fly rapidly away to the north, and then he drove on.

The next day O'Barski returned to the site, parked his car, and walked up the rise into the field, where he saw fifteen little holes, each about four to five inches deep. The holes are part of a chain of evidence that Hopkins and his friend Ted

Bloecher assembled through a combination of diligence and good fortune. Although there were no holes in the turf when Bloecher, Hopkins, and O'Barski returned to the scene nine months after the sighting, there were fifteen little circles of dirt. Hopkins located the park custodian who told him that he had filled in the holes at the sighting location early that summer.

O'Barski's sighting took place next to the Stonehenge Apartments, a cylindrical high-rise overlooking the Hudson River to the east of the park. The Stonehenge lobby faced the field where O'Barski saw the ETV, and Hopkins realized that the doorman on duty that night must have seen what O'Barski saw. Hopkins had sold a large painting to a Stonehenge Apartment tenant in 1968. He located and interviewed the doorman, who remembered helping Hopkins deliver the painting to the tenant. The doorman said that on the night of the O'Barski sighting, at about two or three in the morning, he saw a row of ten or fifteen regularly spaced bright lights shining down from the park.

He could make out a dark mass surrounding the lights. He walked over to the lobby window for a better view. As he called to alert one of the tenants to the sight, he heard a high-pitched vibration and then a sudden crack: the lobby window had broken. When he looked up again, the lights were gone. He immediately called the police about the cracked window, but he did not tell them about the lights in the park: he knew they would understand vandalism but not UFOs. The doorman later met with Hopkins, Bloecher, and O'Barski and walked them into the park to show them where he had seen the lights. It was the same place O'Barski had seen the craft and the occupants, and where there had been fifteen holes in

71

the grass. These are the main links in a chain of additional evidence and witnesses assembled by Hopkins and Bloecher.

In 1976 Hopkins wrote an article about O'Barski's experience for *The Village Voice* (Figure 28). The article was reprinted in *Cosmopolitan* magazine a few months later. People who met Hopkins in art galleries started to tell him about their ETV sightings and close encounters. The O'Barski story thrust Budd Hopkins into his parallel careers of painter and sculptor, and ETV and abduction researcher (32, 33) .

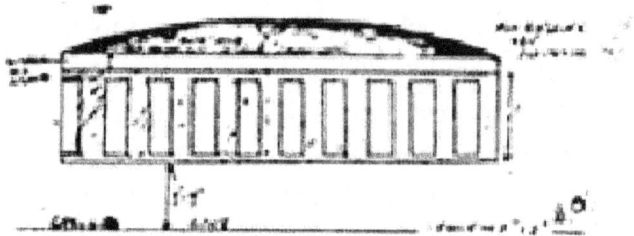

Figure 28. Sketch from the *Village Voice* of the ETV that landed occupants in North Hudson Park

Harare, Zimbabwe; 1994

Emily Trim, the eight-year old daughter of a Canadian Salvation Army couple, was attending the Ariel School in Ruwa, Zimbabwe on September 16, 1994. Morning recess was called, and while the staff attended a meeting inside the school, the children went outside to play. A small ETV landed at the edge of the large playground and two ETs got out. Emily and another girl. Lisil Field, were nearby. They suddenly came face to face to two beings, who communicated messages about the environment and technology, all this was done telepathically.

The school bell rang and the ETs and the ETV suddenly disappeared. Many of the other children saw the ETV and the ETs. The ETV sighting and landing was publicized locally and eventually internationally. It drew attention from John E Mack, a Harvard Medical School psychiatrist whose book, *Abductions*, was published in 1994. Emily Trim returned to live in Canada and Lisil Field now lives in the United Kingdom.

Many of the Ariel school children drew pictures of what they had seen. Emily, as an adult, has kept on drawing and painting. Many of her drawings and paintings are based on her Zimbabwe experience. My wife and I met Emily in the early 2000's and we saw a collection of the paintings that she had assembled in Montreal. Her story has become better-known since it has been discussed at ET conferences and through *The Phenomenon,* a documentary film released in 2020 (34, 35).[1] Here is Emily's drawing of the ETs she saw in Zimbabwe (Figure 29).

---

[1] The official documentary film regarding the event in Zimbabwe will be launched soon under the name "Ariel School Phenomenon," produced by Randall Nickerson

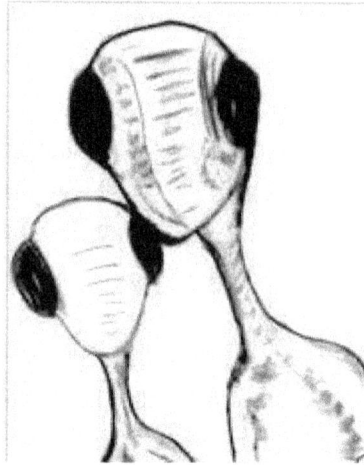

Figure 29. Emily Trim's drawing of ETs she saw in Harare, Zimbabwe in 1994

The ET evidence is human experience that contradicts theory

The evidence reported in *Truth* is based on human experience. And, acknowledging Samuel Johnson again, this human experience contradicts theory for two reasons: it describes vehicles (ETVs) that we do not know how to make, and it introduces us to beings (ETs) whose origin and biology we do not understand. Here is a summary of the ET evidence.

Miriam Bush, a RAAF hospital administrator, saw recovered ETs lying on hospital gurneys. She was quoted in the Roswell incident history written by Thomas Carey and Donald Schmitt (25). Other witnesses who saw the ETs included civilian and military personnel involved in handling and transporting the remains of the crashed ETV and the dead ETs. The multiple-witness accounts of the 1947 crash and recovery of an ETV and ETs are consistent.

Police officer Lonnie Zamorra saw two ETs standing near a landed ETV in Socorro, NM in 1964. Zamorra and another officer saw the ET take off. The ET left indentations and scorched foliage on the landing site that were seen and photographed by witnesses.

Willie Begay and Guy Tossie's car was driven off the road and brought to a halt by a hovering ETV in 1967. An ET levitated down from the ETV and entered their car. The emotional state of two men after the close encounter was described by people who saw them soon afterwards, including the local deputy sheriff and highway patrol officers. Another witness saw an ETV zig-zagging at a low level in the same area at about the same time.

George O'Barski saw an ETV in 1974. It landed near his car in a park, late at night, in North Bergen, NJ, while he was driving home from work. He saw ETs digging in the ground near the landed ETV. The ETV landing was observed by a nearby witness and the holes dug by the ETs were later found and filled by a park employee.

Two schoolgirls (along with many others who were not interviewed) were playing outdoors during recess in Harare, Zimbabwe in 1994. An ETV landed next to the field, and the two girls, who were playing near to where it landed, got a close-up look at two ETs who got out of their landed ETV to look at them. Then the ETs re-entered the ETV and it flew away. I have met both of the now-adult witnesses.

An *Alien Discussions* conference was held at the

Massachusetts Institute of Technology (MIT) in 1991. The conference proceedings were published in 1994. They were edited by David E. Pritchard, an MIT physics professor, and John E. Mack, the Harvard psychiatrist who published *Abductions* in 1994 (36). The conference was about ETs, not about ETVs; the participants (including me) took ETVs for granted.

The UFO Evidence, Volume II (2001) is a 600-page sourcebook of information about ETVs and ETs It was published by NICAP (the National Investigations Committee on Aerial Phenomena) and edited by Richard Hall. It includes many ETV reports and also contains 107 ET reports dating from 1954 through 1989 (37).

The evidence cited in these two books suggests that we have already encountered several different varieties of ET. The most frequently encountered are the so-called short greys, like the three-to-four foot ETs found after the Roswell Crash (Figure 25) and the ETs seen by Lonnie Zamorra near Socorro, NM (Figure 26). ETs with reptilian-appearing skin, like the ones seen by Willie Begay and Guy Tossie in Ririe, Idaho (Figure 27), and by Emily Trim in Harare, Zimbabwe (Figure 29), have also been reported. ETs reported by witnesses whose testimony was presented at the *Alien Discussions* conference, and ETs reported to the many therapists who currently counsel people who experience ET abductions, include a variety of the standard "grey aliens" in sizes ranging from about three and one-half feet tall to six feet tall or more, with complexions ranging from greenish-gray through bluish-gray to plain gray, as well as creatures who looked more like bipedal insects – and even some that looked like human beings. I have no idea how to explain the genetic, linguistic, political or cultural differences or interactions among these ETs, and nothing that I have read about their differences or interactions rises above speculation. [2]

5. Abductions

Extraterrestrials routinely kidnap people. They use ETVs to hunt us from above and they use telepathy to neutralize our resistance. They drag the neutralized abductees into a landed ETV or they beam them up into a hovering ETV (Yes, just like *Star Trek*. ETs know how to overcome the effect of gravity on a large ETV and they know how to overcome the effect of gravity on a small human). Once abducted, ETs examine us. Sometimes they interfere with the reproductive systems of men and women and produce hybrid children by artificial insemination. Inseminated female abductees bring their hybrid fetuses almost to term on earth; then the pregnant abductees are again abducted and the hybrids are delivered aboard the ETV. The hybrids have a variety of fates which will be discussed in more detail later.

The evidence about abductions, like the evidence about ETVs, tells us that we have lost control of our destiny. That aspect of our future will be discussed in more detail in the *ETs.* section at the end of the book. Here are four well-documented abduction cases.

---

[2] My colleagues on the Mutual UFO Network (MUFON) Experiencer Resource Team collect and document much of this information.

Barney and Betty Hill

Barney Hill was a World War II army veteran. He met Betty Barrett while on vacation in New Hampshire. Betty and Barney (Figure 30) married in 1960 and lived in Portsmouth, New Hampshire. In September 1961 they took their first vacation together. They drove west with their dog Delsey to Niagara Falls, then crossed into Canada and drove to Toronto. Then they drove to Montreal where they planned to stay the night. But Barney took a wrong turn, they found themselves south of the city, and they decided to drive back into the United States before stopping.

Figure 30. Betty and Barney Hill

Barney and Betty crossed into Vermont on September 19 and drove east, stopping to eat at a restaurant in Colebrook, New Hampshire. They left the restaurant just after 10 p.m. and drove south through northern New Hampshire on US Route 3. Knowing that traffic would be light, Barney told Betty that they would be home in Portsmouth by 2:30 or 3:00 a.m. The moon was approaching full in the southern sky, Jupiter was visible to the left of it, and Saturn was visible just below the moon.

Driving south of Lancaster, New Hampshire, Betty noticed a light in the sky above Jupiter. The light was bigger than the planet and it seemed to be moving. Betty nudged Barney, who slowed the car to get a better look at it. He told her that it must be a satellite. The light was sometimes invisible as the road turned and trees or mountains obstructed their view, so they found it hard to tell whether or not it was moving. Delsey began to get restless; she needed to go outside. When they came to a stretch of road with good visibility, Barney pulled over, stopped the car and Betty took the dog out. Now that they were stopped, they could see that the light was moving. They looked through binoculars and saw more than a light.

They drove on. The object played tag with them as they drove along Route 3 through the valley of Franconia Notch and past what was then the granite outline forming the Old Man of the Mountain (then New Hampshire's state symbol, it fell off Cannon Mountain in 2003). They thought that what they were seeing might be a commercial airliner *en route* to Canada, but the theory was disproved by the object's sudden halts and turns. Then they thought that a light plane pilot was having fun with them, but when they stopped to look and listen they heard no engine. South of another rock feature called Indian Head; the object hovered about one hundred feet over the road in front of their car.

It was a disc with a row of brightly lit rectangular windows, sixty to eighty feet in diameter and about twenty feet high.

Barney stopped the car in the middle of the road while the disc moved silently to the left and came to rest over the tree line just past the edge of a field. Barney got out of the car, bringing his binoculars with him, and walked across the field toward the disc.

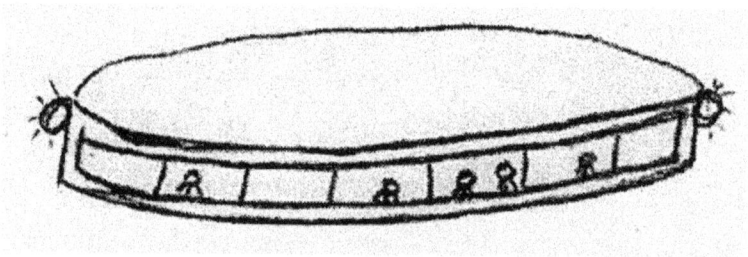

Figure 31. ETV seen by Barney and Betty Hill as it approached their car.

Looking through the binoculars he saw a crew of black-clad figures lined up at the windows Suddenly all but one of them turned away and moved toward what appeared to be an instrument panel. One figure remained looking out the window, and Barney felt that he was about to be captured. Betty screamed at him to get back in the car. Barney tore the binoculars from his eyes, raced back to the car, and they drove away. As they fled, the disc moved overhead and paced them down the road. They heard rhythmic buzzing tones coming from the car trunk. Then they heard a second set of tones. Sometime between hearing the two sets of tones, Barney and Betty had unclear memories of seeing a roadblock and a fiery globe resting on the ground. They continued along Route 3, looking in vain for an open restaurant, until they reached Interstate 93 at Ashland, and then the Route 4 turnoff to Portsmouth. They got home just after 5:00 a.m., as light streaked the eastern sky—later than the 2:30 to 3:00 a.m. arrival that Barney had predicted.

Barney and Betty were subdued when they arrived home. They retrieved their luggage from the car and went to sleep. When they awoke, Barney suggested that they separately draw a picture of what they had seen. They did, and the drawings were similar (Barney's is shown in Figure 31).

Betty called her sister Janet to describe what she and Barney had seen. Janet had seen an ETV in Kingston, New Hampshire (near the *Incident at Exeter* sightings described earlier) and she suggested that Betty report their sighting to Pease Air Force Base, which she did. Meanwhile, Betty had discovered some highly polished round spots on the trunk of their car. Acting on the advice of one of Janet's physicist friends, Betty held a compass over the spots and watched as the compass needle spun round and round. It terrified her.

Betty forwarded an outline of their experience to the National Investigations Committee on Aerial Phenomena (NICAP) which, as you learned from the Ririe, Idaho case, was cautiously interested in "occupants." Walter Webb, a local NICAP investigator, contacted Betty and interviewed the couple in their home toward the end of October. Webb wrote a report to NICAP. Two IBM engineers who had heard about the case through NICAP also wrote to the Hills and asked to talk to them. At the end of November, the Hills, the IBM engineers, and a retired Air Force intelligence major who was a friend of the Hills discussed the encounter in what turned out to be a five-hour meeting. The engineers noticed something that Barney and Betty already knew—they had arrived home much later than they should have, even accounting for the stop-and-go driving as they tried to figure out what they were seeing near Franconia Notch.

About ten nights after their sighting, Betty had nightmares that lasted for five straight nights. The nightmares traced a continuous sequence of events that started with a conscious recollection of the first set of tones until the time when, following the second set of tones, their memory returned, they regained Route 3 and continued to drive south. The nightmares were detailed and fearful blends of her conscious memories and unfamiliar images: being stopped on a secondary road and escorted aboard a landed ETV, being subjected to tests, experiencing unexpected pain, being relieved of the pain, talking to some of the ETs, and being escorted back to the car with Barney and watching as the ETV took off and flew away. A friend suggested that writing down the nightmares might relieve the anxiety Betty felt when she remembered them, so she wrote them down and put the notes away.

Two years after the sighting, Barney had an emotional flashback about what happened at the roadblock that he and Betty had vaguely recalled seeing. Barney was with his family at the time, including Betty's niece Kathleen Marden (an ET investigator and colleague), and as he was recounting the events of the sighting he suddenly cried out in fear as he remembered seeing ETs in the road, who signaled him to stop the car by swinging their arms like pendulums. Then the car motor died and the ETs began walking toward the car with a swaying gait. Barney's fear was so strong that his family remembered his reaction years afterwards.

People suggested that hypnosis might remove the amnesia for

events that Barney and Betty experienced between the two sets of tones. Barney also had chronic high blood pressure and stomach ulcers that were not responding to treatment, and his physician thought that the symptoms might be psychosomatic. He recommended that Barney see a psychiatrist. With Barney's health and Barney and Betty's shared amnesia in mind, the Hills were referred to Dr. Benjamin Simon, a Boston psychiatrist who used medical hypnosis to help World War II veterans suffering from combat stress (now called post-traumatic stress disorder). Hypnotic recall enabled the hypnotized soldiers to remember and re-live their traumatic experiences, thus expressing and purging the debilitating fear that was associated with them.

The Hills' hypnosis-based therapy began near the end of 1963, about two years after their ETV encounter on Route 3. Dr. Simon treated each of them separately in a soundproof room and tape- recorded each session. He spent several weeks training them to enter and leave a deep hypnotic trance on command. He hypnotized them to help them recall what happened during their amnesic period, and then he used post-hypnotic suggestion to prevent them from remembering what they had just recalled under hypnosis (this suggestion was removed at the end of the therapy). This meant that Barney and Betty could not talk to each other about their recalled memories as the therapy progressed. Dr. Simon conducted ten separate hypnotic sessions with Barney and Betty during the first few months of 1964, concluding with post-hypnotic follow-ups that were finished by the end of June of that year.

Barney and Betty's memories, recalled separately under deep hypnosis, were like Betty's nightmares. Betty remembered becoming semiconscious as the first beeping sounds began, and then trying to regain awareness after Barney suddenly drove their car off Route 3 and onto a smaller road. She willed herself back into consciousness and recalled that they were each escorted out of their stopped car by a separate group of two or three ETs. The ETs told her they just wanted to do some tests, and that they would return them to their car and let them go on their way as soon as they were finished. Betty found that her captors were too strong for her to resist so she eventually gave in, entered the disc by an oval door, and was taken down a corridor into a triangular-shaped room.

Barney was brought into the disc behind Betty. He had not regained full consciousness and was carried into the disc dragging his feet; he remembered his shoe-tops scuffing along the ground as he was half-carried and half-floated into the craft. Betty saw that he was taken to a separate room. The ETs removed Barney and Betty's outer clothes. An ET damaged the zipper on the back of Betty's dress and she had to finish unzipping it. Both Barney and Betty were examined by one ET who moved between two rooms, examining each of them in turn. Barney remembered a rectal probe and a groin probe and a genital/urological probe. The examiner used needles connected to wires to probe Barney's skin, and then examined and took samples from his ears and skin. The ETs inspected Barney and Betty's mouths and were surprised to find that Barney had removable teeth (dentures) while Betty did not. Barney also remembered looking into the examiner's mouth as the ET was bending over him, and seeing a membrane that moved when the examiner made mmm-mmm-mmm sounds that seemed to be directed toward other ETs in the room.

There was another ET in Betty's examination room whom she called the "leader." Although he made sounds that were not

English words, she understood them as if they were English. She could also understand some of what the examiner said in the same way. After the leader told her that it would not hurt, the examiner inserted a large needle into Betty's navel. It hurt badly, apparently surprising both the examiner and the leader. The leader waved his hand over Betty's face, and the pain went away.

When the examiner left the room to examine Barney, Betty was left alone with the leader. She felt more at ease since the examination was over, and they had a conversation. Betty said she was astonished at their encounter and opined that it would be hard to convince anyone that it had happened. She said that she would like to take a souvenir with them, and the leader looked around and asked her what she would like to take. She saw something like a book lying on a cabinet and asked if she could take it. The leader asked if she could read it. She said she couldn't, but that did not matter; the book was simply a proof of their meeting. The leader agreed.

Then she asked him where they were from. He asked her if she knew much about the universe, and after she told him what very little she did know, he opened a star map. Betty recalled afterward that it was perhaps three feet wide, oblong, and almost self-luminous, like a holographic projection. It showed bright circles of various sizes connected by solid lines between the larger circles, and some dotted lines. She was told that the circles represented stars or planets, the lines represented trade routes, and the dotted lines represented expeditions. The leader asked her if she knew where she was on this map, and she said no. At which point the leader telepathically said, "If you don't know where you are, there wouldn't be any point in me telling you where I am," and closed the map.

Barney was escorted back to their car while Betty was talking to the leader. As she was preparing to leave, a commotion arose among the crew. A single ET who was shorter, rounder-faced, and apparently angrier than the others, and who seemed to Betty to be in a position of authority, objected to giving Betty the book, so she was forced to give it back. The leader also said they had decided to prevent Barney and Betty from remembering what had happened to them. The leader said even if Betty remembered anything Barney would not, or if he did, his memories would contradict hers and they would end up totally confused.

Betty returned to the car where Barney was already sitting behind the wheel, and where Delsey the dog had been left during the abduction. They sat in the car and watched the disc take on an orange glow, lift into the sky, and disappear from view. Barney started the car, maneuvered back down the dirt road and rejoined the concrete pavement of Route 3. Then the second set of beeping tones occurred, and they regained full memory near Interstate 93 at Ashland. They drove down I-93 until they reached Route 4, turned east on Route 4 and drove home to Portsmouth.

Dr. Simon never endorsed the reality of Barney and Betty Hill's abduction experience, but he did admit that he could not explain much of what the Hills told him under hypnosis. He did not know where Betty's nightmares came from, nor did he have an explanation for why they matched her recovered memories. Dr. Simon was skeptical, however, of Barney's story. He knew that Betty had told Barney about her nightmares and suggested that Barney was "filling in" with information he had remembered hearing from Betty.

Although the therapy led to partial relief of Barney's symptoms, the Hills were disappointed that Dr. Simon would not say that the experience was real. Nevertheless, Barney and Betty remained in touch with him and he remained willing, when given the Hills' permission, to talk about their experiences and his role in the case. Dr. Simon did believe that Barney and Betty had seen an ETV.

Physical evidence supports Barney and Betty's narrative. First is the set of polished circles on the trunk of the Hills' car that made a compass needle spin when placed on top of them. The spots gradually faded away after a year but were seen by many witnesses, including Betty's niece Kathleen Marden. The Hills believed these spots were associated with the two sets of tones they heard at the beginning and end of their amnesic period.

Second, there is a broken binocular strap. Barney remembers tearing the binoculars from his eyes and racing back to the car when he felt that they were about to be captured; the strap probably broke then.

Third, both Barney and Betty's mechanical wristwatches stopped working after the trip and could never be made to work again.

Fourth, there is Betty's dress. When the Hills got back from their trip, Betty took the dress off, hung it in a closet, and never wore it again. When the dress was inspected later, the lining was ripped and there were tears in the threads holding the zipper in place. This damage is consistent with the memory that an ET could not manipulate the zipper, got it stuck, and

Betty had to finish unzipping it.

Fifth, Barney's shoe-tops were scuffed, as if they had been dragged along the ground. They would not scuff on top as a result of normal walking. The scuffed shoe-tops are consistent with Betty's memory that Barney was dragged into the landed disc by two ETs.

Sixth, after the sighting, Barney developed a circle of twenty-one three-quarter-inch warts on his groin. He remembered under hypnosis that some kind of probe or instrument had been placed against his skin. They were not venereal warts and in fact seemed to be a benign if ugly growth, and Barney eventually had them removed.

The seventh, and certainly the strangest, piece of physical evidence was the return of Betty's earrings. A few weeks after their encounter, the Hills locked up their house and left on a day trip up Route 3 to look for the landing site. They returned that evening, unlocked the house, and found a pile of dried leaves on the kitchen table. The earrings that Betty had worn on the September journey were in the middle of the pile. She had missed them but had not done anything about it.

Dr. Simon reminded them that if they had been abducted, it must have happened somewhere. Although they had not found the location on their first few day trips, eventually they turned off Route 3 and crossed a trestle bridge that they remembered seeing on the night of the abduction, turned up a narrow dirt track off the secondary road, and came to a clearing that they both recalled as being the abduction site.

The Interrupted Journey (38) was the first book about their experiences, and it was written by John L. Fuller, the author of *Incident at Exeter*. A second book: *Captured: the Barney and Betty Hill UFO Experience* (39), was written many years later by Kathleen Marden, Betty Hill's niece, and Stanton Friedman, a well-known ET researcher, who was also a college classmate and a friend of mine.

Two of Betty's remembered experiences were followed up by additional investigations. First, Betty remembered under hypnosis the star map she had been shown by the leader, and her post-hypnotic drawing was published in *The Interrupted Journey*. Amateur astronomer Marjorie Fish, a teacher and MENSA member, saw Betty's drawing and created a three-dimensional model to learn whether Betty's remembered map matched any location in our milky way galaxy. She located all of the stars within fifty-five light-years of the sun by reviewing printed star catalogs, and she used different sizes and colors of beads to represent different types of stars. The beads were suspended using nylon fishing line. Her three-dimensional construction began in 1968 and evolved as new and more accurate star data was obtained and published. In 1972, using data from the most recently published catalog, she found a single position from which the Betty Hill two-dimensional star map superimposed itself accurately on the stars in her three-dimensional map. She identified the two main stars of Betty Hill's remembered map as Zeta 1 and Zeta 2 Reticuli, and found that one of the stars joined to the pair of main stars by lines in the Betty Hill map was our sun (Figure 32).

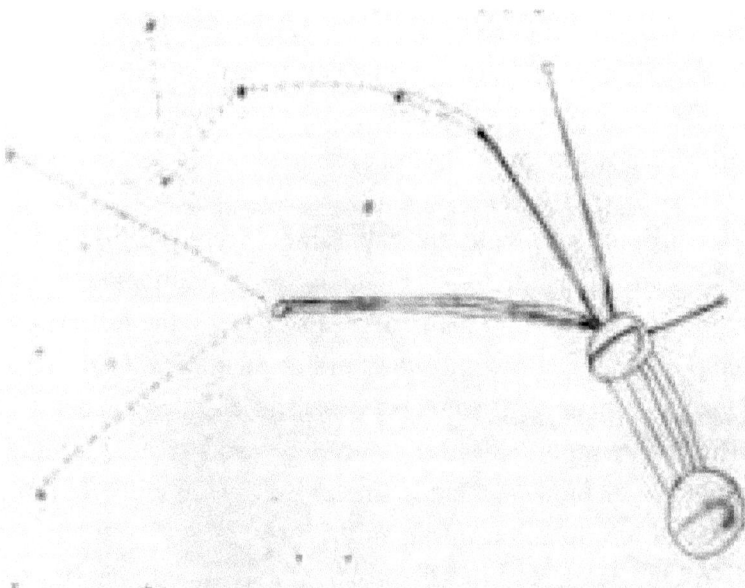

Figure 32. The star map Betty remembered seeing aboard the ETV.

Second, Betty remembered that during her abduction she had been offered a souvenir book by the leader, but another ET overruled that decision and Betty was forced to leave the craft without the book. Betty later drew the symbols she remembered seeing on the book, and in 2007 her drawing was published for the first time in *Captured!* Budd Hopkins had collected drawings over many years from abductees who said they remembered seeing symbols inside a craft. The symbols Betty remembered and that were published for the first time in 2007, look remarkably like many of the symbols that Hopkins had been collecting from abductees since 1975 (Figure 33). Symbols seen inside ETVs have been remembered by other abductees and the similarities among remembered symbols will be discussed in more detail in Chapter 6.

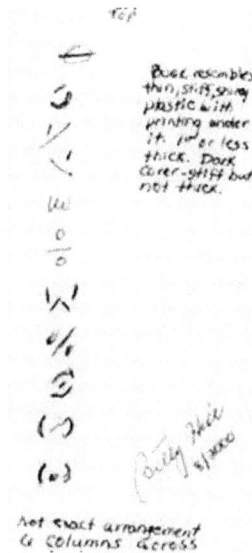

Figure 33. Symbols on the book Betty was not allowed to take from the ETV, remembered under hypnosis.

The Abduction Revolution

The Hills' abduction was the first widely reported case of humans being kidnapped by ETs, examined aboard an ETV and returned to earth. If it had remained an isolated case, it would just be an anomaly: the remembered experience, physical evidence, nightmares, flashbacks, and hypnotically recalled memories a joke arranged by Coincidence to baffle Humans. But it was not an isolated case: it was the documented precursor to many others.

Here is a summary of the historical revolution – a turning point for humanity – that was first widely reported with the Hill abduction:

- A close encounter with an ETV.

- Missing time.

- Physical evidence of an abduction: for example,   the magnetized spots on the car, scuffed shoes, torn clothing, and Barney's groin warts.

- Nightmares and flashbacks about the abduction experience.

- Hypnotically recalled memories of the abduction experience that are consistent with the nightmares, flashbacks, and missing time.

If abductions are a consistent experience and not an anomaly that common sense tells us to ignore, there should be more accounts like Barney's and Betty's, reported by reliable observers and analyzed by competent researchers. There are.

Michael Lapp and Janet Cornell at Buff Ledge, Vermont

Two camp counselors were abducted from the Buff Ledge Camp on the eastern shore of Lake Champlain near Winooski, Vermont, on the evening of August 7, 1968. Most of the campers and staff had gone to an event in nearby Burlington and the waterfront was deserted. Two of the counselors, Michael Lapp and Janet Cornell (not their real names), put on bathing suits and went down to the dock to talk and swim.

At about 8:10 pm, just after sunset, they noticed a bright object in the southwestern sky. It swung down in a long arc to the right, moving about fifty degrees in a few seconds. When it stopped it looked large, cigar-shaped, and self-luminous. Three small lights emerged from the end of the cigar-shaped ETV. The lights carried out some visually compelling maneuvers, spiraling upward, fluttering downward, and eventually appearing as luminous discs as they moved closer to Michael and Janet. Two of the disks departed abruptly and the third disk, making a sound like many tuning forks, plunged into the lake, creating whitecaps and wind that persisted until it rose out of the water a few minutes later and hovered over the dock where Michael and Janet stood. The disc was "as big as a small house"— Michael estimated that it was forty to fifty feet in diameter—and had a transparent dome on top through which they could see two occupants (but bigger than the Ririe, Idaho ETV shown in Figure 27).

Michael turned to the figures in the ETV and said out loud: "What do you want? Where are you from? Are you going to hurt us?" and to his surprise a voice answered (in his head): "We are not here to harm you." Michael and the occupant carried out a conversation in which he spoke but the occupant's part was "telepathic." Janet remained incommunicado.

The ETV hovered over the dock and shone down a bright light that enveloped Michael and Janet. Michael lost consciousness. When he regained consciousness, it was about 9:00 p.m. and he heard the sound of car doors slamming as the campers and staff returned from Burlington. Two of the returning campers saw lights over the dock. They ran down to the water's edge and saw the ETV disappear into the night. A staff member saw the ETV from an upstairs window of the camp's main building.

Neither Michael nor Janet ever publicized their close encounter with an ETV and ETs. Years later Michael, who still lived in Vermont, decided to explore the experience he had never forgotten and had never resolved. He called the Center for UFO Studies (CUFOS, mentioned in connection with the Taconic Parkway sightings, pp 18–20). CUFOS referred him to Walter Webb, the NICAP field investigator who first interviewed Barney and Betty Hill. Webb, an astronomer, was a senior lecturer, assistant director, and operations manager of the Charles Hayden Planetarium at the Museum of Science in Boston.

Webb invited Michael to Boston for an interview. Lapp remembered witness Janet Cornell and the other campers who saw the ETV, but where were they now? Lapp and Janet had not been in contact since the summer of 1968. Michael found Janet's current business address, Webb wrote her a letter, and she called him back. Michael and Janet agreed to be hypnotized to try and learn what had happened during their "missing time." Although Janet lived in Atlanta, she returned routinely to the Boston area on business so it was possible to arrange hypnosis sessions with Boston-area hypnotherapists who volunteered their services. Michael was interviewed a total of five times by two therapists and Janet was interviewed separately, twice, by one therapist. During this time Michael and Janet met once, for two hours at Boston's Logan Airport while Janet was waiting for a connecting flight. Michael had completed one hypnosis session by then and Janet had not yet started hers, but Webb, who was with them, made sure nothing that Michael described during his hypnosis session was discussed at their meeting. Webb was present at all the interviews. Unlike Dr. Simon and the Hills, the therapists did not block Michael or Janet's later recall of what they remembered during hypnosis.

Michael and Janet's conscious and hypnotically recalled memories agree that they were standing on the dock directly under the ETV when an intense beam of white light shone down on them. Michael remembered pushing Janet and himself to the ground, and that is where their conscious memories ended. Under hypnosis, they remembered being floated up the beam of light and into the transparent dome of the ETV, where they joined four occupants. The occupants were short, had greenish-blue skin, large heads, thin necks, large, goggly, protruding eyes, small or rudimentary ears and nose, and small mouths, and their hands appeared to have three long fingers. They appeared to be wearing skin-tight, silver-gray clothing. They communicated with each other through a high-pitched, continuous sound but communicated with Michael and Janet telepathically, mostly to reassure them that they would not be hurt. Michael and Janet were examined on a tilting examination table. Michael remembers little of his own exam but remembers seeing Janet's examination, part of which she also remembers. He and she both remember probes being placed on various parts of their bodies.

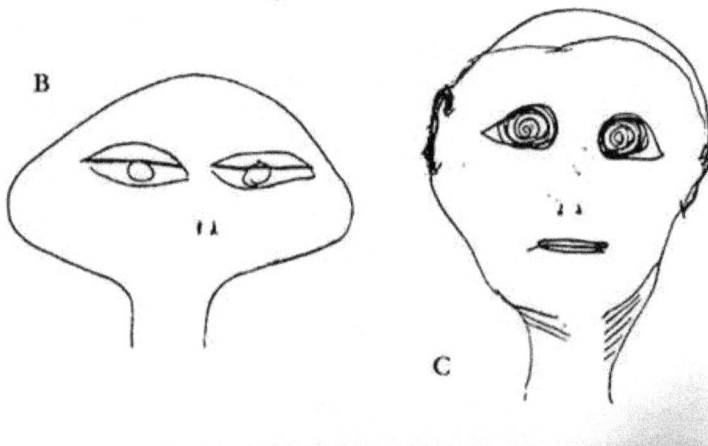

Figure 34. Michael Lapp's sketch of the Buff Ledge ETs

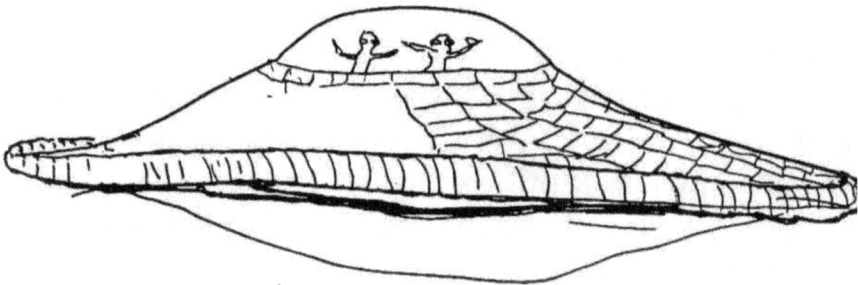

Figure 35. The ETV that hovered over the dock at the Buff Ledge camp.

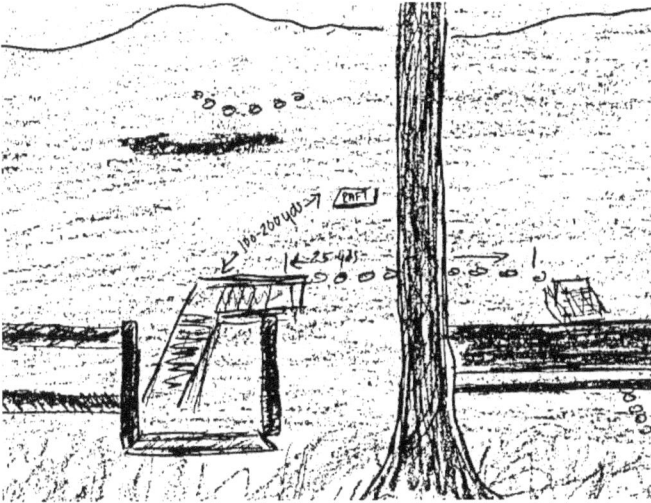

Figure 36. Drawing made by a Buff Ledge witness who saw the ETV hovering over the dock before it disappeared.

Michael remembered that the ETV over the dock rendezvoused with a very large ship, possibly the cigar-shaped one they had seen first. When he looked through the transparent dome as they approached the large ship, he could see the Earth the size of "a nickel at arm's length" and the moon to the side. Webb estimated that the observation point might have been about 200,000 miles from Earth. The smaller ETV entered a hangar inside the larger ship and Michael and Janet floated down a "light tunnel" that joined their smaller ship to a door in the hangar wall.

Janet had far less detailed memories of the larger ship, but Michael remembered a confusing scenario of humans, some clothed, some naked, congregating in a misty outdoors environment. His memories of what followed were not resolved, but he also remembered seeing display screens that presented rapidly changing information continuously, and he was told that he was learning it for use in the future. Finally, he remembers seeing an image projected on the screens of he and Janet lying on the dock, at which point his conscious memory returned and he found himself, with Janet, back on the dock.

After following many leads, writing many letters, and making many phone calls, Webb was able to find all of the witnesses that Michael remembered as well as several more ex-campers and counselors who also saw the mysterious lights over the dock. Figures 34 and 35 are drawings made by Michael Lapp following his hypnosis sessions, and Figure 36 was drawn by a returning witness who saw the ETV depart. They were published in Webb's 1994 book, *Encounter at Buff Ledge* (40). By then Michael Lapp had graduated from college with a bachelor's degree. Janet Cornell, three years older than Lapp, had graduated from college, earned a masters' degree *summa cum laude*, found professional work, become the wife of a physician, and then a mother.

The Barney and Betty Hill abduction and the Buff Ledge abduction were investigated by professionals (psychotherapists and hypnotherapists) who have helped interested amateurs to study ETVs and ETs. I had no direct involvement in either case, although I am a research colleague of Kathleen Marden, Betty Hill's niece, who grew up with the Barney and Betty Hill experience as part of her family (and her own) history. The next accounts are classic but also more personal because the investigators are people I know or knew well, and I have met the abductees.

Linda Cortile

I met Linda Cortile (not her real name, Figure 37), under conditions that, as you will learn, convinced me that her abduction experience was traumatic and real. This account is based on my discussions with Linda and with other witnesses to her abduction, on a review of some of the original documents related to her experience, and on the lucid description of her experiences that was recorded by the late Budd Hopkins in his book: *Witnessed: The True Story of the Brooklyn Bridge Abductions* (41).

Figure 37. Linda Cortile

Linda Cortile is the first-generation daughter of Italian immigrants, and had lived in New York City all her life. She had read an earlier book by Budd Hopkins about abductions that had taken place in Indiana. The last page of that book explained how to report an abduction experience by contacting Hopkins through his publisher. In April 1989 she wrote to Hopkins describing experiences that had begun in her parents' home while she was living there as a young adult, and that persisted after she married and left home. Hopkins, a well-known abstract expressionist painter, also lived in New York City, about a mile and a half from Linda's apartment.

In May 1989 Linda met with Hopkins to discuss her experiences. When she was eight years old she remembered seeing a brightly-lit object hovering over a nearby apartment building. Hopkins induced hypnosis, and Linda recalled more clearly that the object was cone-shaped and small—not more than about fifteen feet high and ten feet wide. Hopkins had planned to schedule more hypnotic sessions to explore various aspects of Linda's earlier experiences, but that schedule was pre-empted when Linda called Hopkins on December 1, 1989, distraught by what she remembered from the night before.

On November 30 Linda went to bed late, long after her husband was asleep (he worked an early shift). As she got into bed at 3:15 a.m., she began to feel numb. She tried and failed to wake up her husband. She saw something hiding behind the window drapes. The drapes parted, and a short, gray creature approached the bed. She threw a pillow at it and suddenly felt that she had done something terribly wrong. She had a fragmentary memory of a piece of white gauze floating in front of her face, and of sitting on a table while someone thumped lightly up and down her back. That was all she could consciously remember. This memory frightened and depressed her because she thought her unusual and upsetting experiences were behind her.

At about 3:00 a.m. on the morning of November 30, a VIP

security convoy started to drive south from midtown Manhattan on the FDR Drive along the East River. The convoy turned onto South Street, just before the drive becomes an elevated highway. (Security convoys avoid elevated highways because they might be trapped and ambushed there.) The convoy was heading to Battery Park, on the southern tip of Manhattan, to meet a ferry that would carry their VIP to a heliport on Governor's Island. US security agent Dan, driving, was accompanied by agent Richard in the passenger seat. Their VIP was in the back seat. For no apparent reason the motors in the convoy cut out, and the convoy came to a stop on South Street near the Brooklyn Bridge and about five hundred feet from Linda's apartment building. Dan and Richard tried to radio agents in other cars, but the radio did not work. They asked the VIP to lie down in the backseat: the VIP took the unexpected turn of events good-humoredly and complied. The agents decided to do nothing for a few minutes and see if the sudden motor and radio failure would correct itself before taking the potentially risky step of leaving the car to get help.

Richard reached into his pocket for a stick of gum, and as he unwrapped it an orange glow reflected off the foil wrapper. He looked up to see what was causing the glow and saw a reddish-orange EIV hovering over a nearby apartment building. Richard, Dan, and the VIP saw the ETV move away from the top and then down the side of the building until it hovered just above an apartment window. A beam of light shone down from the ETV. A small creature floated out of the apartment window into the beam. The creature was followed out the window by a woman dressed in a white nightgown that floated up over her head. Two other creatures followed the woman out of the window. They all remained suspended briefly in the light beam before floating up into the bottom of the ETV. The light beam disappeared and the ETV flew over the Brooklyn Bridge to the south. They watched as it plunged into the East River just a few hundred yards beyond the bridge.

Dan and Richard had also read Hopkins' earlier book about the Indiana abductions. They wrote to Hopkins, describing what they saw. Their letter, as Hopkins put it, "drove a spike straight through the heart of reason." They had seen Linda Cortile's abduction. .

On the same night, Janet Kimball (not her real name) was returning from Brooklyn to upstate New York. At some time after 3:00 a.m. she was halfway across the Brooklyn Bridge when her car lights dimmed and her motor cut out, as did the lights and motors of the other cars on the bridge. She looked to her right and saw an ETV hovering near an apartment building close to the bridge. She saw a small creature suspended above an apartment window, a woman below the creature, and two creatures below the woman. She saw them rise into the bottom of the ETV and saw the ETV fly directly over the bridge, where it disappeared from view behind the bridge's elevated walkway.

Months later, Hopkins received a thick envelope containing Janet Kimball's drawings of her experience. The drawings were like Richard's drawing illustrating what he had seen of Linda Cortile's levitation into the ETV. Janet Kimball had seen the same thing Dan and Richard saw: Linda Cortile's abduction (Figures 38 and 39).

I met Linda Cortile while visiting Hopkins in 1992. I sat

through one of the hypnosis sessions described in Hopkins' book, *Witnessed* (39). Linda was reclining on a couch. I sat seven or eight feet beyond Linda's feet. Hopkins began to induce hypnosis and I began to doze. Hopkins asked Linda to remember what happened in her bedroom. She said she decided to check out something suspicions at the bedroom window curtains: "I just get up from bed and I walk to the window . . . *ooohhhhh!*"(bloodcurdling scream). That high-decibel scream snapped me out of my doze and made me appreciate the emotional impact of a remembered abduction experience. And that afternoon I drove Linda home from Hopkins' apartment in the Chelsea neighborhood.to her apartment building next to the Brooklyn Bridge,

I also met another witness at Hopkins' house (not named in his book) who happened to be looking out of her apartment overlooking the East River, late at night, on November 30, 1989. She also saw the ETV hovering over Linda Cortile's apartment building near the Brooklyn Bridge on the night of the abduction.

Hopkins received an unsolicited letter from the VIP that secret agents Richard and Dan were protecting (which I have also seen). In it, the VIP acknowledged that he had shared the experience with Richard and Dan, whom he knew had been in contact with Hopkins. He apologized for not having contacted Hopkins sooner, marveled about the significance of the experience, and hoped that both he and Hopkins would enjoy reminiscing about it in their old age: "Won't it be amusing to sit back and watch all the nations of the world pull together. It shan't be long before the earth becomes whole again." But the diplomat refused to publicly discuss his involvement with the experience, and signed the letter "The

Third and Last Man." Hopkins knew who the VIP was, and so will everyone else who knows anything about international politics and who reads *Witnessed*.

Figure 38. Agent Richard's sketch of Linda Cortile's abduction

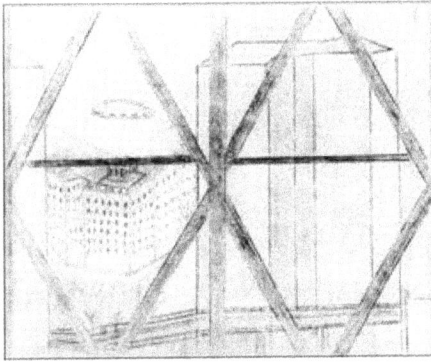

Figure 39. Janice Kimball' sketch of Linda Cortile's abduction, seen from the Brooklyn Bridge

Sean Allen

Sean Allen (again, a pseudonym) was well-known to David Jacobs, a leading investigator of the abduction phenomenon. David spent his professional career at the History Department of Temple University in Philadelphia, Pennsylvania where, along with other subjects, he taught a one-semester course called "Unidentified Flying Objects in American Society." David's first book was his PhD thesis: *The UFO Controversy in America* (42). He has written several other ETV-related books including the one that describes Sean Allen's experiences: *Walking Among Us: The Alien Plan to Control Humanity* (43). I met David years ago at an ET conference in Philadelphia. I met Sean Allen, an English businessman, in 2015 in Oxford, England.

Younghae Chi is a member of the Faculty of Oriental Studies at Oxford University and is a friend of David and of me.

Younghae does not list ETVs and abductions as one of his interests on the Oxford faculty's website, but he is interested and he has co-authored a book (in Korean) about ETVs and extraterrestrials. In the spring of 1915, at Younghae's invitation, I came to Oxford's Nuffield College to talk about ETVs and extraterrestrials. The evening after the lecture, Younghae drove his wife, my wife and I to a country inn outside of Oxford. There we met Sean Allen and his friend Joanna (also a pseudonym). We had an engaging and pleasant evening.

Several days later (Friday, May 8: V-E day) my wife and I, back in London, joined the crowd in Parliament Square to see the fly-by, to listen to the drums and bugles of the Grenadier Guards Band, and to watch the Queen, followed by the then mayor of London (as I write, Prime Minister) Boris Johnson, enter Westminster Abbey for the V-E day celebrations. Then, by pre-arrangement, we crossed the Lambeth Bridge to the Thames Embankment (the south side of the Thames, from which London is in view) and met Sean and Joanna. They drove up in Sean's Jaguar.

We walked and we talked, ate lunch and dinner together, and rode the London Eye, a giant Ferris Wheel,. Sean owns a company that manufactures precision medical instruments that are used around the world, and he travels widely to represent his company and its products. He has had close encounters with ETVs, and has experienced multiple abductions by ETs, ever since he was a boy living in northern England. He told us about an abduction that happened while he was on a business trip in Istanbul; during that abduction there was an entity accompanying the ETs who looked like his grandfather.

Remember as you read this that 'Sean Allen' is a technically educated and successful businessman, that he and his company have an international clientele that he visits regularly and an internet presence under his real name. His life is a counter-example to skeptics who think that abduction narratives are driven by character weakness and psychological needs (more about that in Chapter 6). Sean's story, like Linda Cortile's, drives, as Budd Hopkins put it, "a spike straight through the heart of reason."

Sean was once abducted while on a trip to the United States, and was given a job to do by his abductors. He was introduced to a group of normal-looking young men and women aboard the ETV. But they weren't human. These human-looking 'hubrids,' as he called them, had been conceived by women who had been abducted, inseminated with genetically modified sperm harvested from abducted men, and returned to earth. When the women were close to term they were re-abducted, the developing fetuses were delivered aboard the ETVs, and the fetuses were brought to term in artificial placentas aboard the ETV, an idea first proposed as a means of "improving" the human race by Aldous Huxley in *Brave New World* (44). The genetically modified sperm had been altered by the ETs to create 'hubrids' who look like humans but who can communicate telepathically. Hubrids use telepathy to make humans do what the hubrids want them to do.

Hubrids are destined for life on earth; their job here is to blend into human society and to occupy an ordinary place in the everyday life of the communities where they are introduced. But there is a problem: they lack many human social skills. Although raised aboard the ETV and tutored in basic human behavior (language, hygiene, manners) they lack refinements that only contact with human society can provide. Sean Allen was recruited to teach them social dancing!

Jacob's 2015 book describes in greater detail, with many more examples, "the alien plan to control humanity" (his book's subtitle), and it explains how groups of 'hubrids' have taken up residence on earth, found employment, and have been trained to blend in. One story, coming from one person, about something so unlikely, is reasonably regarded with caution if not suspicion when first heard. But it is neither the first nor the only story. Nor will it be the last.

Hybrids

Three new realities have been presented in this book, in decreasing order of everyday experience and in increasing order of significance to us. The first is that ETVs are extraterrestrial vehicles. The second is that extraterrestrial beings (ETs) crew the ETVs. The third is that humans are abducted into ETVs by ETs. The even more disturbing fourth reality, introduced by Sean Allen's story, is that ETs inseminate abducted human women to produce a hybrid species of human and extraterrestrial, some of whom live among us on earth. Here are two more accounts that support that reality.

Kathy Davis

Kathy Davis (not her real name) was abducted into ETVs several times from her home in the American Midwest. During one of the abductions, she was inseminated aboard the ETV. During a second abduction her developing fetus was extracted and brought to full term aboard the ETV. At the time of the first abduction Kathy was engaged to be married and she thought that she had become pregnant by intercourse with her fiancé. After the second abduction, her period returned and all the signs of her pregnancy disappeared, leaving her saddened and upset by the loss. On a subsequent abduction she was introduced to her artificially inseminated hybrid daughter aboard an ETV.

Kathy, along with some of her friends, had experienced more than one consciously remembered close encounter with ETVs. Fleeting memories of close encounters, plus an inability to remember in detail the course of events following the encounters, is called "Missing time," a possible indicator of a suppressed abduction experience. *Missing Time* was also the title of Budd Hopkins' first book about ETVs (45). Hopkins was also the author of *Witnessed*, the book about Linda Cortile's abduction (41). Kathy Davis contacted Hopkins because she came across *Missing Time* and because her remembered experiences were similar to those reported in the book. And in due course, Linda Cortile contacted Hopkins because she had read *Intruders* (46), his book about Kathy Davis' abduction.

Hopkins had learned the skill of hypnosis when he started to investigate the ETV abduction experience, because so many people who consciously recalled seeing an ETV could not recall what happened during the time immediately following the abduction. Their memories were either gone or suppressed. Hopkins thought those probably traumatic memories had been repressed, and so have many other students of trauma, starting with Sigmund Freud (47). Kathy Davis came to New York at Hopkins' invitation. Hopkins conducted many interview sessions with Kathy Davis, which outlined the series of events described briefly above. His hypnosis sessions also revealed subsequent abductions, during one of which Kathy was presented to Emily, the hybrid offspring of her earlier insemination, subsequent abduction and fetus removal (Figure 40). Emily, as far as we know, still lives with the extraterrestrials aboard an ETV.

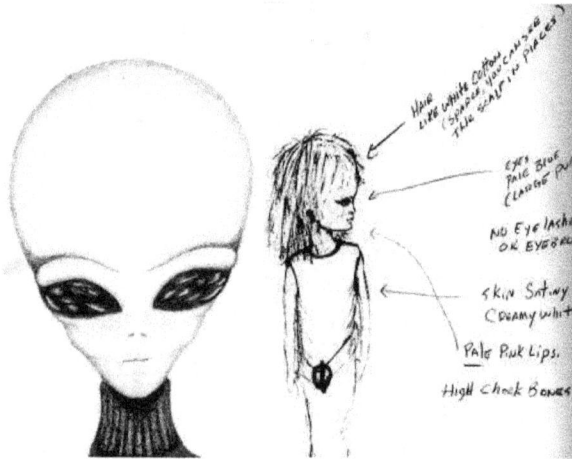

Figure 40. Drawing of an ET as remembered by Kathy Davis, and a drawing of her hybrid daughter Emily who, as far as is known, still lives aboard an ETV.

Sylvie P.

Sylvie P. is a French-speaking Canadian who is the mother of hybrids. *Les enfants de Sylvie P. (The Children of Sylvie P.),* the book describing her experiences, was written by Pierre Caron and Marc St-Germain (48). I have met Sylvie, and I wrote the introduction to the book. The book describes Sylvie's encounters with extraterrestrials and how those encounters became the central feature of her life. What happened to Sylvie is like what happened to Kathy Davis as told by Budd Hopkins in *Intruders*.

As a child Sylvie P. lived with her parents and a younger sister and brothers at the end of a country road in the hilly Laurentian Region of Quebec. Her first clearly remembered encounter with an ETV occurred just after supper on an early summer evening when she was about 13 years old. She and her sister were throwing a ball back and forth outside when they spotted a luminous ball of white light hovering over some nearby electric wires. They were frightened. As the sphere moved away they ran inside. One of their brothers said that the house lights had just dimmed. The fright passed and the sisters decided to go outside and play again, but their mother told them it was too late: it was daylight when they had gone outside to play but it was past 10 pm when they ran in to tell the family what they had seen. Sylvie P. and her sister experienced what Budd Hopkins called "missing time."

Sylvie P.'s remembered close encounters with ETVs, and her disturbing but incomplete memories after each encounter led her in 2016 to seek out someone who could help her to clarify her ETV experiences. She composed a handwritten letter, accompanied by sketches of the remembered close encounter, to Marc St-Germain, the Quebec chapter director of MUFON (The Mutual UFO Network, an international ET research organization to which I belong). Marc enlisted the help of Pierre Caron, a Quebec-based hypnotherapist. Pierre carried out six hypnotic regression sessions with Sylvie during 2016 and each session was recorded. The memories Sylvie recalled during the hypnosis were not hidden from her by a suggestion to forget (as Benjamin Simon had done with Barney and Betty

Hill, see p. 74). The earth-based part of her ETV story happened in rural Quebec, but her experiences aboard ETVs are familiar from the accounts of abductees everywhere.

Over the course of the six hypnotic regression sessions, Sylvie revealed that she had been abducted routinely from the time she was a child. At one point, when she was about 12, she was abducted and inseminated, but the fetus did not live. The ETs showed up in a field where she was picking raspberries, moved her to an abandoned building, and extracted and removed the stillborn fetus. At 15, Sylvie was again abducted and inseminated aboard an ETV. About eight weeks later, she was re-abducted and her developing fetus was removed and incubated to term aboard the ETV. Then the abduction, insemination and removal process happened again. The result of these events, which included the insemination and removal of a stillborn fetus and the insemination and removal of live fetuses brought to term aboard an ETV, was that Sylvie suffered from anemia as well as other gynecological conditions that left her incapable, as an adult, of normal conception and pregnancy.

Sylvie P had contact with ETs from early childhood until just a few years before she finally sought help and contacted Marc St-Germain in 2016. The disturbing history of her repeated abduction and sexual exploitation by ETs is narrated through the six detailed accounts of her hypnosis sessions with Pierre Caron and Marc St-Germain that are reported in *Les enfants de Sylvie P*.

6. Who gets abducted?

Abduction and hybrid reports are the most bizarre stories about people that you have ever read. Not the worst stories – even the abductees who were sexually exploited are here, albeit with medical and psychological trauma, to tell us about it. All of these reports are based on fleeting, incomplete memories that were recorded after an interaction between the person who had the experience and a researcher or clinician who often used hypnosis as a tool. Did these people suffer from personality defects that explain their stories? Or did the hypnotist suggest the stories to them? The answer to both questions is no. After the ETV evidence was presented, you read about experiments and patents that helped to explain how ETVs might work. Now that the abduction and hybrid evidence has been presented, here are some clinical studies and experiments that explain why the abduction and hybrid reports are true.

Abductee personality

Do personality disorders explain the experiences reported by abductees? Dr. Aphrodite Clamar, a clinical psychologist, interviewed nine abductees ranging in age from twenty-eight to forty-three. They had various white-collar occupations including secretary, audio technician, business executive, and corporate lawyer. Five were men and four were women. Four were unmarried, one was married, and four were divorced. They were all college-educated and two had graduate degrees. Four of them had reported serial abductions and several others were thought to have had serial abductions. Their abductions were reported from as early as five to seven years old (like

Sylvie P), and they had reported abductions until as recently as three years before the study began. On the basis of her interviews, Dr. Clamar found no reason to believe that their abduction reports were caused by personality disorders.

In order to confirm her own assessment, Dr. Clamar contacted another clinical psychologist who did not know the nine abductees and was not told, either by Dr. Clamar or by the abductees, about their abduction experiences. The second psychologist, Dr. Elizabeth Slater, was hired to give each of the nine people a regular psychological assessment which included an interview and a battery of psychological tests. She was told by Dr. Clamar that her research group was studying "creativity" and they were interested in learning about the similarities and differences among the nine study participants.

Dr. Slater based her analysis on the nine abductees' responses to the Wechsler Adult Intelligence Scale (WAIS), the Rorschach test, the Thematic Apperception Test (TAT) and the Minnesota Multiphasic Personality Inventory (MMPI). These are standard tests routinely used by clinical psychologists. She also interviewed the nine abductees, and she based her written conclusions on both the test results and on the interviews. She wrote that the outward personalities of the nine people ranged from exhibitionistic and dramatic to shy and reserved, and that their predominant abilities ranged from verbal to practical and mechanical. Despite these overt differences, she found that their underlying emotional makeup was similar. They all suffered from anxiety and were uncomfortable with other people. Dr. Slater reported that the nine abductees were all above average in intelligence and one was in the top range on the intelligence test. She wrote that none of them suffered from personality disorders.

Dr. Clamar then hired a *third* clinical psychologist, who did not know either the abductees or their common abduction experiences. She asked the third psychologist to review and comment on Dr. Slater's test results and interview reports from five of the nine abductees, as well as on her summary report. The third psychologist was satisfied that the diagnoses and conclusions of those reports were consistent with Dr. Slater's interview notes and the psychological test results.

After Dr. Slater's reports had been reviewed by the third psychologist, Dr. Clamar told her about the common experiences of her nine clients. Dr. Clamar wrote that "[Dr. Slater] was, it is safe to say, flabbergasted." Dr. Slater then wrote this comment in a supplemental report. "The first and most critical question is whether our subjects' reported experiences could be accounted for strictly on the basis of psychopathology, i.e., mental disorder. The answer is a firm no." Dr. Slater wrote that her subjects were not pathological liars, paranoid schizophrenics, or dissociative personalities, and that nothing in their personality profiles suggested that they would either confabulate such experiences or lie about them. She did note that their anxiety, inner turmoil, weak sense of identity, and suspiciousness of others, while not proof (an after-the-fact correlation is never proof), was at least consistent with having experienced the powerlessness and trauma associated with the reported abduction experiences (49,50).

The American Personality Inventory

Budd Hopkins worked with Ted Davis, a psychologist and social worker, to develop a test to determine whether someone had been abducted by ETs. Psychologists like long tests: the

Minnesota Multiphasic Personality Inventory (MMPI, mentioned above) has over five hundred true-false questions. The Hopkins-Davis Personality Test, as it was called, had 608 true-false questions. The test made no mention of ETs, ETVs or abductions. The 608 questions were all about experiences and attitudes within the range of normal life that Hopkins and Davis thought might give answers pointing to an abduction experiences without giving away the purpose of the test.

They gave the 608-question test to 52 people (26 men and 26 women) who reported ET abduction experiences and who had contacted Hopkins and Davis or other American investigators about their experiences. Twenty of the ET abduction group had spontaneously recalled being abducted by ETs while 32 others had undergone one or more regressive hypnosis sessions that established their abduction experiences before they took the test. The investigators had agreed that the abduction experiences of all 52 people were real. They also gave the test to 22 "controls:" people whom they had no reason to believe had ever experienced an ET abduction.

After collecting the 52 abductee and 22 control test results, Hopkins and Davis asked me to help them "validate" the test, which means to show that the test can distinguish between people with self-reported or hypnotically elicited abduction experiences, and people who had never reported an abduction. It was useful that I lived in Montreal, Canada, far from the NYC Metropolitan area where Hopkins and Davis were both known as abduction researchers. It was also helpful that as a university professor I had access to both a pool of test subjects and to undergraduate research assistants who could help me to carry out the study. Given that Budd Hopkins was well-known both as an ET investigator and as a painter and sculptor,

I asked if we could change the name of the test to conceal its purpose. They agreed, and it is now called the 'American Personality Inventory' (API).

The 608 API questions were transferred to digital format so the test could be administered on laptop computers. Research assistants were enlisted to administer the API to two groups of volunteer subjects recruited in Montreal. (I thank Amanda Baker, Heather Pulman, Tamara Lagrandeur, Julia Bellissimo, Brian King and Erin Friend for collecting most of the data) One subject group was told that we were developing a new personality test and that they were being asked to participate as a "control group" sample of psychologically normal people. This group included both university undergraduates and the parents of undergraduates. Adult test-takers were included to make the sample more representative of the people who report abductions. The second group, recruited in the same way, was treated very differently. We first asked them what they knew about the ET abduction experience. After they explained what they did know, and after having confirmed that none of them had really experienced an ET abduction, the researchers asked this group to answer the API questions "as if" they had been abducted by ETs. This was our "simulator" group.

The details of the analysis are complex but are fully explained in the published research paper.. Using a statistical tool called "canonical discriminant analysis" we found that responses to a set of 65 of the 608 true-false questions of the API allowed us to separate the 52 abductees tested by Hopkins and Davis, the 75 "controls" tested in New York and in Montreal, and the 26 "Simulators" tested in Montreal. The result is shown graphically in Figure 41. Each axis represents one of two variables that were calculated from responses to 65 of the 608 questions. Figure 41 shows that the ET abductees do not answer those 65 questions like the controls, who we have no reason to believe have ever been abducted, nor like the simulators, who are people who we asked to "fake" having

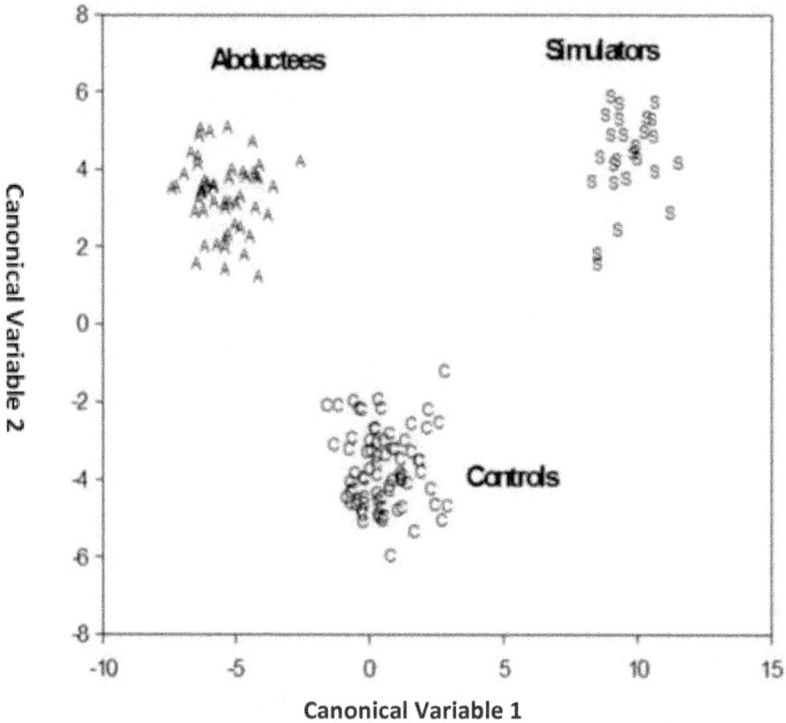

Figure 41. Separation on two variables (vertical and horizontal axes) of API respondents in the abductee, control and simulator groups tested in the Davis, Donderi and Hopkins study (2013)

been abducted by ETs. We conclude that the ET abduction experience is real, that it leads to experiences that are captured by answers to 65 questions on the API, and that the people who report real ET abductions can be easily separated on the

API from people who have not had this experience and from people who pretend to have had this experience (49).

The API Eleven Years Later

The internet, as used by MUFON, The Mutual UFO Network, gave the API extended usefulness as a tool to study ET abductions. People who contact MUFON about ET abduction experiences are routinely asked to complete an online questionnaire called the Experiencer Survey (ES). Starting in January 2018 they were also asked to complete the 65-question version of the API. The results from 175 of the people who completed the API have been analyzed. The position of each of the 175 new MUFON respondents was plotted, along with the results from the original 153 respondents, on the graph shown in Figure 42. There is greater variation among the 175 MUFON respondents along both dimensions than was true of the original 52 abductees tested by Hopkins and Davis, but the *average position* of the 175 MUFON respondents (the blue dot) on both dimensions is at the center of the cluster of ET abductee results collected by Hopkins and Davis and analyzed using the same statistical method. Despite the greater variability of the MUFON respondents, the 175 people who reported abduction experiences to MUFON gave API results that were on average like the 52 Hopkins-Davis abductees and not at all like the controls or the simulators who were tested in the original Donderi, Davis and Hopkins study (51).

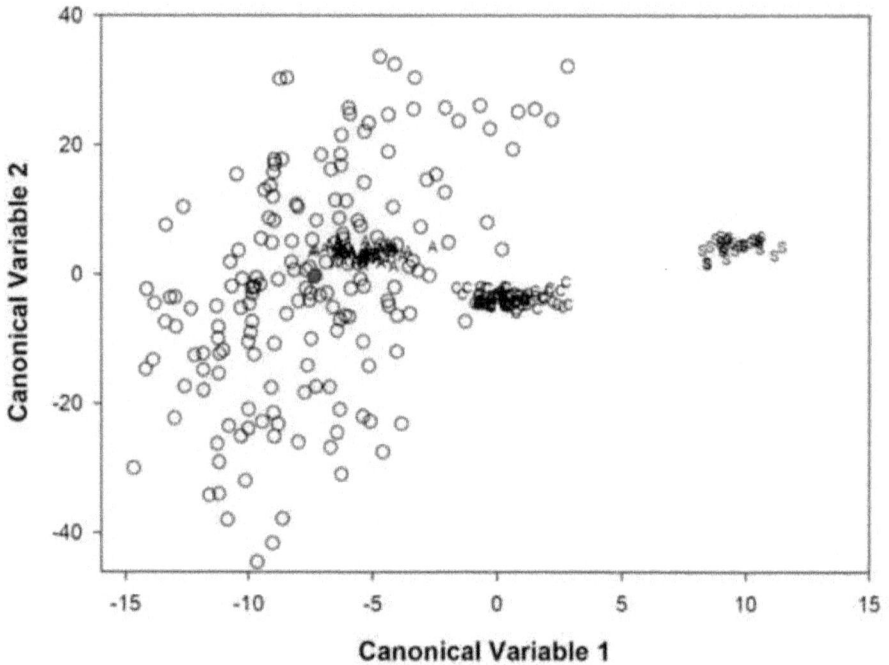

Figure 42. The API test (65 questions) as answered by 175 people from the MUFON experiencer group, plotted in blue on the same graph with the original abductee, control and simulator groups (plotted in black) from the Davis, Donderi and Hopkins (2013) study.

Abduction Symbols, Hypnosis and Reality

Many of Budd Hopkins' abductees recalled under hypnosis that they had seen symbols of one sort or another inside an ETV (as did Betty Hill, see figure 33). Hopkins asked his abductees to draw the symbols that they remembered, and he kept the drawings. Stuart Appelle, like me, was an experimental psychologist interested in the ET phenomenon. Stuart was also a hypnotist. Knowing that Hopkins had a collection of symbols that were remembered under hypnosis by people claiming to have been abducted, Appelle tested the idea that symbols remembered by "abductees" had nothing to do with abductions and everything to do with suggestibility under hypnosis. So he hypnotized 24 university student volunteers, none of whom said that they had experienced an abduction. Appelle asked each of these non-abducted volunteers, under hypnosis, to experience an "alien abduction." And after the "abduction experience" was over, Appelle asked his volunteer subjects to draw the "alien symbols" that they recalled when they were "abducted." He kept the drawings.

Now there are two sets of "alien symbols" remembered under hypnosis. One set was drawn by fourteen people who Budd Hopkins thought had really been abducted by ETs. The other set was drawn by 24 people who Stuart Appelle had no reason to think had ever been abducted by ETs. If the symbols in the Hopkins "real abduction" set look like the symbols in the Appelle "imaginary abduction" set then there is no reason to think that the "real" abductions are any more real than the "imaginary" abductions. The conclusion would be that the whole abduction experience is imaginary.

I had developed a testing method that makes it easy for observers to evaluate the similarities and differences among sets of things like pictures. The method makes it possible to answer the question: do the "real abductee" symbols look like the "imaginary abductee" symbols, or are they different?

Nineteen experimental participants who knew nothing about and were told nothing about any of the symbols were simply asked to judge the similarities among the entire set of symbols obtained from both Hopkins' "abductees" and Appelle's "imaginary abductees." Each symbol was presented by itself on a neutral grey card, and the symbol cards were mixed up randomly and then spread out on a large table. Each of the experimental participants was individually asked to sort the symbol cards into groups. They were asked to put symbols that looked similar into the same group. They were allowed to use as few or as many groups as the liked, so long as there was more than one group and there were fewer groups than the total number of symbol cards.

If the judges mixed up the "real abduction" symbols with the "imaginary abduction" symbols, that would suggest that the abduction symbols remembered by Hopkins' abductees were just as imaginary as the "abduction" symbols remembered by Appelle's imaginary abductees, and that would suggest that the abduction experience reported by Hopkins' abductees was as imaginary as the abduction experience that Appelle's subjects had been hypnotized and then asked to imagine.

This experimental method allows for failure because it allows for a so-called "null hypothesis," or negative outcome. The "null hypothesis" would be that there was no difference found between the symbols remembered by Hopkins' 24 "real" abductees and the symbols remembered by Appelle's 19 "imaginary" abductees. This would mean that both sets of symbols were equally imaginary and therefore, so were the "abductions" falsely remembered by Hopkins' 24 "abductees."

It did not turn out that way. Instead, the symbols remembered by Hopkins' abductees were judged to be different from the symbols produced by Appelle's imaginary abductees on three different psychological dimensions (Figure 43). This noticeable (and "statistically significant") difference between the symbols drawn by people who reported an ET abduction, and the symbols drawn by people who imagined an ET abduction, tells us that the memories of people reporting an ET abduction are different from the memories of people asked to imagine an ET abduction, and this supports my conclusion that the ET abduction experience is real (52, 53, 54).

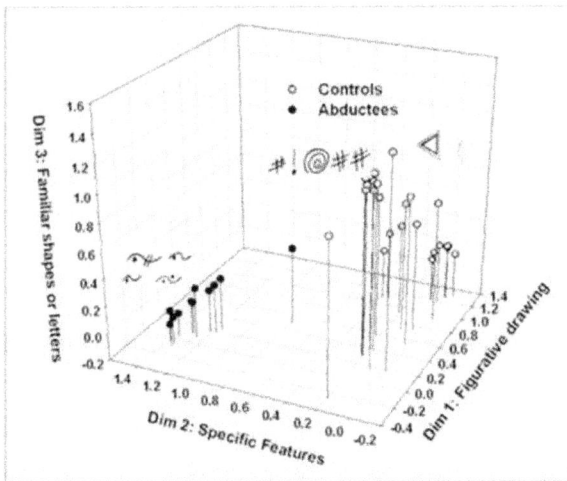

Figure 43. Differences on three psychological dimensions among the remembered symbols seen by Budd Hopkins' abductees (black symbols) and the symbols created by Appelle's simulated "abductees" (open symbols). Representative symbols from each group are shown on the graph.

The Abduction Poll

Many more people know about ETVs than know about the ET abductions described in this book. But more people may have been abducted than we might at first think. In 1991 about seven thousand randomly selected Americans were interviewed in order to estimate how many of them had been abducted by ETs. They obviously weren't asked "Have you been abducted by ETs?" Instead, they were asked indirect questions that were included among many other questions asked during a routine door-to-door commercial Roper Poll. The Roper poll asked each respondent many questions about things of interest to its commercial sponsors (cars, appliances, etc.) It also included questions paid for by Robert Bigelow, a supporter of ETV research. The questions were chosen by abduction researchers Budd Hopkins, David Jacobs and Ron Westrum. The yes-no questions asked about "unusual personal experiences" that earlier studies had shown might be consciously remembered by abductees, even though the abductee might have been telepathically told by the ETs to forget the actual abduction event. None of the questions mentioned ETVs, ETs or abductions. Five of the questions were: whether the person remembered waking up paralyzed with the sense of a strange figure in the room, remembered experiencing missing time, remembered a feeling of flying, remembered balls of light in the room, or had puzzling scars following the experience of missing time. About two percent of the poll sample, representative of all adult Americans, answered "yes" to four of those five questions, leading the researchers to conclude that two percent of adult Americans, about 3.7 million people in 1991, had been abducted (55).

Telepathy is the ET Abduction Tool

ETs transmit thoughts, feelings or ideas to people without using visible symbols or audible speech: in other words, they communicate with us by telepathy. Telepathy experiments already show that people can communicate with each other without auditory or visual contact and that people can accurately describe details about things that are far away without knowing anything in advance about those things (56,57,58). Although we have been studying telepathy for a long time, we are just now beginning to learn how to control it. Over the last fifty years the CIA has spent time and money studying and using "remote viewing:" a form of extra-sensory perception, to help them learn about things when a direct approach would be either physically or politically impossible (59). In 1979, and without the help of the CIA's budget, then graduate student (now MD) Howard Eisenberg and I reported a successful telepathy experiment in a peer-reviewed scientific journal (60).

"Brain waves;" which are measures of electrical activity at different parts of the surface of the brain, can be recorded from electrodes placed on the outside of the head. This is called electroencephalography.. Electroencephalography was discovered in 1929 and is now used routinely in medicine. A recent experiment shows that when your "brain waves" are recorded and then transmitted directly to someone else's head, you can communicate actions to that person, who can be far away from you: you wave your hand or tap your foot; your "brain waves" reflect the action, and the person receiving *your* brain waves on the outside of *his* head can recognize your actions: the hand wave or the foot tap, and respond accordingly. This is a far cry from the success that ETs have with telepathy, but we are just beginning to learn how to exploit EEG signals to generate telepathy (61).

The ETs are ahead of us. Almost every interaction between ETs and humans is telepathic. Although abductees sometimes hear sounds emanating from ETs, they do not understand the sounds as speech. The Barney and Betty Hill abduction, the Michael Lapp and Janet Cornell Buff Ledge abduction and the Linda Cortile abduction, all described earlier, show that the abductee knows what the ET wants him or her to do, and then does it, but the abductee does not hear human speech from the ET.

Electromagnetic telepathy

Inventors act where theorists fear to tread. People standing near radar antennas sometimes hear "clicks" as the electromagnetic radar signals are pulsed out. Based on that observation, a patented invention claims that messages might also be directly conveyed by electromagnetic radiation. The human skull (with your brain inside) is a tool that converts external messages that you cannot hear (electromagnetic radiation or high-frequency sounds) into internal messages that you can understand. That is one definition of telepathy. Here is the invention that claims to do that (Figure 44):

United States Patent [19]

Stocklin

[11] Patent Number: **4,858,612**

[45] Date of Patent: **Aug. 22, 1989**

[54] HEARING DEVICE

[76] Inventor: Philip L. Stocklin, P.O. Box 2111, Satellite Beach, Fla. 32937

OTHER PUBLICATIONS

Gerkin, G., "Electroencephalography & Clinical Neurophysiology", vol. 135, No. 6, Dec. 1973, pp. 652-653.
Frye et al., "Science", vol. 181, Jul. 27, 1973, pp.

Figure 44. An invention that claims to transmit messages directly to the brain by electromagnetic radiation

The patent description is:

A method and apparatus for simulation of hearing in mammals by introduction of a plurality of microwaves into the region of the auditory cortex is shown and described. A microphone is used to transform sound signals into electrical signals which are in turn analyzed and processed to provide controls for generating a plurality of microwave signals at different frequencies. The multifrequency microwaves are then applied to the brain in the region of the auditory cortex. By this method sounds are perceived by the mammal which are representative of the original sound received by the microphone.

Is the US government interested? Indeed it is. Figure 45 shows that the Small Business Innovation Research Program of the US government has described and funded a US Air Force research program designed to put the Stocklin patent to work.

Communicating Via the Microwave Auditory Effect
Agency:
Department of Defense
Branch:
Air Force
Agency Tracking Number:
19903
Amount:
$739,569.00
Phase:
Phase II
Program:
SBIR

Small Business Information

SCIENCE & ENGINEERING ASSOC., INC.
Sea Plaza, 6100 Uptown Blvd, Ne, Suite 700, Albuquerque, NM, 87110

Figure 45. US Government research program to study electromagnetic telepathy.

Ultrasound telepathy

ET telepathy might also be based ultrasound. Ultrasound is high-frequency sound, beyond the hearing range of people or other animals, which penetrates the skull and, when focused on a specific region of the brain, can modify the function of the neurons (brain cells) in that region. In 2020, a research group from Universities in Utah and California showed that they could force rhesus monkeys to look at either one of two possible visual targets by using ultrasound to stimulate one or the other of two locations in the monkey's brain (62).

We can hear radar clicks. We know how to make monkeys look at one or another of two targets using ultrasound. This is a modest beginning. ETs have been compelling humans to do much more complicated things, without talking to them, for seventy-five years. They might be using electromagnetic radiation. They might be using ultrasound. They might be using something else. We are just beginning to learn how to do the things that ETs already know how to do.

We know the world through evidence before we understand the world by theory. We *do not understand* how ETVs work but we *know* that ETVs are here. We *do not understand* who ETs are, but we *know* that they are here. We *do not understand* how telepathy works, but we *know* that ETs communicate telepathically with abductees.

Protection against ET telepathy

The stereotype of a "UFO believer" is someone who wears a "tinfoil hat" to protect against telepathic control by ETs. By now you should be past thinking that ETVs and ET telepathy are delusions If you are not, please start reading again at page one. What follows is a discussion about how we might prevent what the *evidence* tells us is real: telepathic control of people by ETs. That control may involve radiation – electronic or ultrasound – and this discussion is about how we might prevent electronic or ultrasonic radiation from influencing human behavior.

We now know that non-verbal information can be

communicated either electromagnetically or by high-frequency sound above the threshold of hearing (61,62). Although we do not know how ET telepathy works, our immediate need is to protect ourselves against it. And if ET telepathy is based on electromagnetic or ultrasonic radiation, we know some things that might help.

If emg radiation is responsible for telepathy, then the hat shown in Figure 46 might stop direct emg transmission to your brain and might prevent telepathy from controlling your behavior. Michael Menkin also developed another hat that interferes with emg radiation and so should prevent emg-transmitted telepathy. Menkin's hat uses multiple layers of a commercially available emg shielding product called Velostat. His website explains where to get the materials needed to make the helmet and, once you have the materials, how to assemble it (63). Menkin reports what happened to some 40 abductees who used his hat. For more than half of them, the abductions stopped when they began to wear the hat. Other abductees reported that ETs abducted them if they forgot to wear the hat, or because "hubrids" (p. 107) threatened them physically for wearing the hat and made them take it off.

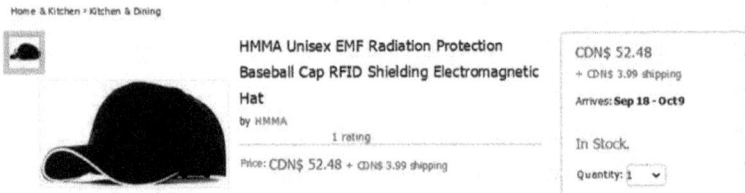

Figure 46. Commercially available EMG protection hat

Protection against ultrasound telepathy

Protection against ultrasound telepathy is also available. A comprehensive US Patent (Figure 47) states that multiple layers of shielding material will protect against ultrasound, and that the shielding "materials can be any materials, including, but not limited to, plastics, polymers, silicones, epoxies, hydrogels, rubber, composites, ...".

US008857438B2

(12) **United States Patent**
Barthe et al.

(10) Patent No.: **US 8,857,438 B2**
(45) Date of Patent: **Oct. 14, 2014**

(54) DEVICES AND METHODS FOR ACOUSTIC SHIELDING

(75) Inventors: **Peter G. Barthe**, Phoenix, AZ (US); **Michael H. Slayton**, Tempe, AZ (US); **Charles D. Emery**, Scottsdale, AZ (US)

4,166,967 A	9/1979	Benes et al.
4,211,948 A	7/1980	Smith et al.
4,211,949 A	7/1980	Brisken et al.
4,213,344 A	7/1980	Rose
4,276,491 A	6/1981	Daniel

(Continued)

Figure 47. Patent recommending multiple polymer layers to shield against acoustic interference.

Velostat, a material described in the acoustic shielding patent, is a flexible composite consisting of a polymer impregnated with carbon black, an electrical conductor. The Menkin hat has similar layers of multiple conduction polymers and it may also block ultrasonic radiation. The Menkin hat, the EMG hat, and the ultrasound patent may be the first steps toward eliminating the threat of ET telepathy from either emg radiation or ultrasound. They may be the first steps towards protecting our species from control and modification by ETs.

7. A Summary of the Truth

The Evidence about the Truth

An expert witness is someone whose training and experience qualifies him or her to provide information that helps a judge or jury to decide a case.[3] The people who flew fighter jets from the *USS Nimitz* and the *USS Theodore Roosevelt* described in *ETs – Yesterday* were aviation expert witnesses. They knew the natural phenomena and the human machines that they could expect to see while flying over the ocean. The "Tic-Tacs" that David Fravor and Alex Dietrich and Ryan Graves and Danny Accoin saw were not part of the natural world or of our man-made world. David Fravor said:

> "It was a real object, it exists and I saw it,"
> Asked what he believes it was, 13 years later,
> he was unequivocal. "Something not from the
> Earth" (64).

Their visual testimony was corroborated by instrumental evidence – radar and infra-red records from aircraft and aircraft carrier sensors. These expert witnesses and the others who reported the same thing saw extra-terrestrial vehicles – ETVs.

Older historical accounts and images have been re-interpreted as UFOs, and there are scattered reports dating from the

---

[3] I spend part of my professional time preparing expert reports about visual perception, and other aspects of psychology, for lawyers arguing legal cases.

nineteenth century, but UFO reports increased rapidly during the twentieth century. Evidence proves that Earth has been visited by *Extraterrestrials*. Almost every ET observation involves witnesses whose reports are corroborated by supplementary evidence. The Roswell ETs were seen and reported by Miriam Bush, a base hospital administrative assistant and they were sketched by Glen Dennis, a Roswell mortician brought to the air force base to deal with the ET bodies. The corroborating evidence is the debris field from the crashed UFO seen by W. W. "Mac" Brazel, who reported it to the Roswell sheriff, who reported it to the Roswell Army Air Force Base. And so began the ET story – and the ET cover-up. The ET explanation of the event was immediately denied by the Air Force. The debris field was explained as the wreckage of a "weather balloon" (July 9, 1947) and fifty years later (June 25, 1997), the recovered bodies were explained as "crash test dummies" (65). The denials continued for 70 years and they will be described in *Lies*.

The ET evidence shows the familiar pattern of eyewitness testimony supported by corroborating evidence. Lonnie Zamorra, a police officer, saw a landed ETV with ETs standing near it in 1968; as he approached, the ETV took off with a fiery roar. Nearby witnesses saw the takeoff and later investigators found indentations and scorched foliage at the landing site. Two ranch workers in Idaho, driving home at night, were forced off the road by a small ETV carrying two ETs who they saw through the transparent dome. They were terrified, one of the witnesses fled the car and pounded on door of a nearby home. The ETV departed. The resident returned with the worker to the car and found the other visibly terrified witness still cowering in the car. A man driving home late at night past a park in North Bergen, NJ, saw a landed

ETV and saw ETs near it, digging in the ground: another witness saw the landed UFO, and as the second witness watched it take off, the window through which he was watching cracked. And in 1994 in Harare, Zimbabwe, a schoolyard full of children playing outdoors at recess saw an ETV land at the edge of their playing field. Here there were no corroborating marks or broken windows but there were scores of eyewitnesses, many of whom were interviewed shortly afterwards and all of whom described seeing the same thing. While this is not "dispositive," to use legal jargon, my wife and I had dinner a few years ago in Canada with Emily Trim, one of the Harare witnesses whose parents, a Salvation Army couple, had gone from Toronto to Zimbabwe to help start the school. She is a credible witness.

Like the evidence for ETVs and UFOs, the evidence for *abductions* starts with the abductee's conscious recall. But this conscious recall is different: it is recall of "missing time" – awareness that there is a gap in the sequence of remembered events from an earlier time to the present. For example: having remembered driving south past a notable landmark on US 3 in New Hampshire, seeing a hovering UFO nearby and then hearing some beeps over your car, your next memory is of being in the same car, hours later and dozens of miles further down the same road (Barney and Betty Hill). Or while sitting on the dock at a deserted summer camp on the edge of Lake Champlain, you see a UFO hovering overhead; your next memory, demonstrably hours later, is of lying on the dock under the hovering UFO as some of the campers, returning late from a nearby town, see the UFO depart as they return (the Buff Ledge abduction). Abductee's "missing time" has been recovered using hypnotic regression: a psychotherapeutic tool used to recover repressed or forgotten memories. It notably did so when a California school bus full of children was captured and hidden by armed kidnappers. The bus driver and children escaped and the bus driver, under hypnosis, remembered the license plate of the kidnappers' car (66). His recovered memory was accurate, and the kidnappers were arrested, convicted and sent to prison.

Sometimes witnesses see the abduction. When Linda Cortile was abducted through her apartment window into a hovering ETV, it was seen and described in drawings by two independent witnesses; one whose car was brought to an involuntary stop while she was driving across the nearby Brooklyn Bridge and the other whose car was brought to an involuntary stop on a nearby highway.

What about the people who report having been abducted? The abductees who have seen psychologists at the request of researchers are not psychologically abnormal, but they are psychologically upset. They can be distinguished from other people, including people asked to fake an abduction experience, by their responses to psychological personality tests. What they remember about their abduction experience under hypnosis is also different from what non-abductees "remember" when asked to imagine an abduction experience under hypnosis. I helped to do some of the research that established the reliability and validity of abduction reports.

Two women mentioned here have reported experiences far more consequential that just an abduction. They claim to have been abducted, inseminated aboard the ETV, returned to

Earth, then re-abducted and delivered of *hybrid* children aboard an ETV. The children looked like crosses between humans and ETs. The evidence includes their conscious recall of a close encounter with an ETV followed by missing time, and then subsequently includes evidence of pregnancy, further abductions and the hypnotically aided recall of seeing and interacting with the hybrid children aboard an ETV.

This is of course a ludicrous confabulation because such things never happen on earth, right? Wrong: we do it all the time. Humans have been selectively breeding other species for centuries: your local dog park shows you the result. We either select breeding partners, or we extract and mix sperm and eggs and then use *in vitro* insemination to create embryos. We do clones and we do hybrids. Evidence shows that for whatever reasons, ETs are hybridizing us.

All of this happens without violence, because ETs use telepathy to persuade us to do what they want us to do. *Telepathy is the ET abduction tool.* We know very little about telepathy (non-verbal, non-instrumental communication at a distance) except that although we have studied it for years and we do know that it works (67), we do not know how to use it as a reliable way to communicate. Research suggests that telepathy may be a form of ether high-frequency ultrasonic or electromagnetic communication directly to the brain, but neither publicly available research nor publicly available theory explains how to use it reliably. ETs use it reliably.

The Implications of The Truth

The truth is that ETs visit earth regularly. Their ETVs (still called UFOs by the press) are often seen by casual observers. I saw one in the night sky over Montreal in 1985, skipping along like the "flying saucers" reported in 1947 by Kenneth Arnold. Newspaper accounts of ETVs seen from the ground or from ships or airplanes are now routine. In the few years before this book was published, a flood of "UFO" sighting stories have been reported in mainstream print and visual media.

I cite references that include thousands of documented observations, corroborated by multiple witnesses or recorded on video or radar, which have been collected, analyzed and summarized in scores of publications by members of the many private research organizations that study the ET phenomenon. People working with private groups have also gathered and summarized the data released by government agencies: notably thousands of heavily redacted documents from the Department of Defense, the CIA and the NSA (National Security Agency) in the United States, as well as documents from GEIPAN (the French Space Agency) in France (69), from the UK Ministry of Defence, and from the Canadian government's "Non-meteorite sighting file."

The truth is that ETs interact with people. ETVs land near or hover over people and vehicles. ETs abduct people while they are out walking, or from their cars, or while they are at home in bed or watching television. They levitate people into an ETV or they instruct people telepathically to approach the ETV. During an abduction the humans are controlled telepathically by the ETs and are often subjected to intrusive medical examination and interference. Men have semen harvested; women are inseminated. The humans are returned to earth with only fleeting memories of what happened. The same person may undergo repeated abductions, with results that have already been described: the production of hybrid ET-human children, some of whom, visually indistinguishable from us, are destined to live among us. ETs subject us to intrusive, and by our standards criminal, behavior. If we do nothing, ET behavior is not likely to change until they have done whatever they have set out to do with and to us. ET behavior might change if we do something to make it change. How we might change ET behavior will be discussed in the *ETs* section of this book.

Accepting this evidence means that we accept that our species is being observed and manipulated by ETs. This presents all of us with an existential problem. It is harder to accept that we have an existential problem than to accept the simpler ideas, discussed in Chapters 1 through 4, that ETVs are real and that they are here. We routinely find excuses to avoid confronting problems when the solution is unclear, the consequences are severe, and a plausible but specious explanation that avoids the problem is available. The higher the emotional cost of acknowledging the problem, the greater the resistance to accepting the evidence. The psychologist Leon Festinger called this mental escape route "cognitive dissonance" (68) and it is part of the psychology of how people think and behave. Cognitive dissonance encourages people to find specious reasons to reject uncomfortable ideas.

Abductions happen in a world that we already know is visited by ETVs and ETs: that evidence is summarized in Chapters 1 through 4. The abduction narratives are consistent across continents, across languages, across investigators, and over time. They are a set of "human experiences" which, as Samuel Johnson wrote, are "constantly contradicting theory." They are "the great test of truth." These experiences tell us that our species is being examined by, experimented with and genetically modified by ETs.

Lies

You can fool some of the people all of the time, and all of the people some of the time, but you cannot fool all of the people all of the time.

P. T. Barnum

The story of Lies has two sides: the Dark Side (why and by whom the lies were created) and the Good Side (who uncovered the truth, and how they did it). Government officials and spokesmen routinely lied about ETVs and ETs, while ETV and ET researchers worked to expose the truth that the liars tried to conceal. Both sides of the story began in 1947, at the beginning of the modern ET era.

8. The Dark Side

Unnecessary Lies

Suppose that some American nuclear missiles became inoperable and a reporter at a press conference asked an air force officer: "what is the state of our nuclear arsenal?" A necessary lie would be to answer, "just fine," while technicians worked frantically to fix the disabled missiles. Part of the US nuclear arsenal was disabled in 1967 by an ETV hovering over the missile silos (70).

An unnecessary lie would be one that *does not* compromise security against threats from other countries but *does* conceal some part of what we know about the universe. This chapter reviews the unnecessary lies that the US government told about ETVs from 1947 through 2017. No other government has gone to such lengths to discredit the ETV evidence.

An explanation for the US government's initial hostility to the ETV evidence is that ETVs were first seen in 1947, at the beginning of the Cold War between the Soviet Union and the western democracies. The American Ground Observer Corps, scanning the skies for hostile aircraft, might mistake ETVs for Soviet bombers, thus accidentally starting a war, or the Ground Observer Corps might be overwhelmed by ETV sightings and let Soviet bombers slip through unobserved. If people were discouraged from reporting ETVs, so the reasoning went, they would be less likely to report "false positives" (ETVs mistaken for Soviet bombers) or to miss a "false negative" (not reporting a real Soviet bomber).

Freedom of thought and expression are protected in the United States and in other western democracies, so the disagreements between governments intent on concealing the ETV evidence and the people and private groups intent on understanding the ETV evidence have been, and still are, played out in public on visual media, at meetings, in books and newspapers and in privately published investigations. Although governments revealed little of what they had learned about ETs and ETVs from 1947 until 2017, people and private groups wrote and talked regularly about what they learned, and they do so to this day. France is an honorable exception to government reticence: its space agency has managed an objective ETV reporting service since 1977 (68).

The Roswell Incident

The lies started suddenly because events happened suddenly during the two weeks between June 24 and July 8, 1947. On June 24, Kenneth Arnold saw "flying saucers" over Washington State, and on July 3, Mac Brazel found the wreckage of a crashed "flying saucer" near Roswell, New Mexico. Both events were described in the *Truth* section of this book and both were reported in newspapers when they happened. Arnold's "flying saucer" account appeared in a widely circulated Associated Press story on June 26 (Figure 48.

PAGE 2   THE CHICAGO SUN, THURSDAY, JUNE 26, 1947

In These United States

Supersonic Flying Saucers Sighted by Idaho Pilot

Speed Estimated at 1,200 Miles an Hour
When Seen 10,000 Feet Up Near Mt. Rainier

Figure 48. The Kenneth Arnold story as reported in the Chicago Sun on June 26, 1947

The Arnold incident was not challenged when it appeared in print, but the US government response to the Roswell incident, about a week later, established a precedent that lasted until 2017. On July 8 the *Roswell Daily Record* printed an accurate account of the Roswell crash. The headline read: "RAAF Captures Flying Saucer On Ranch in Roswell Region." But on July 9 the headline read: "General Ramey Empties Roswell Saucer" (Figure 49). Air Force headquarters provided a photograph of pieces of a damaged weather balloon as an "explanation" for the wreckage that "Mac" Brazel had seen and that the Air Force had retrieved from the Foster Ranch. Starting on July 9, 1947, and continuing until 2017, every US government official quoted or official statement released offered a lie or a misdirection to explain away the recurring ET and ETV evidence.

The Roswell ETV wreckage and the ET bodies were flown to the Air Technical Intelligence Center (ATIC) located at Wright-Patterson Air Force Base in Dayton, Ohio. The Air Force "UFO investigation" started at ATIC immediately afterwards. The investigation never mentioned or dealt with the Roswell wreckage or bodies. The ATIC investigation was called Project Sign in 1948, Project Grudge in 1949, and Project Blue Book in 1952. It ended when Project Blue Book was closed in 1969. That happened after a government-sponsored group called the Condon Committee, described earlier, issued a report that ended any publicly-acknowledged US government interest in extraterrestrials (12).

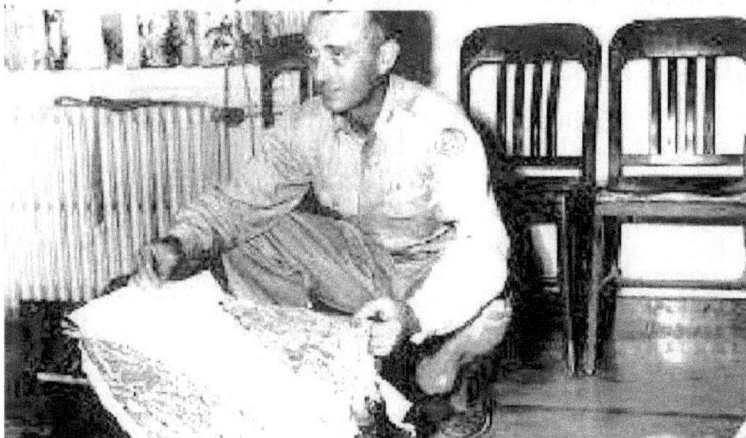

Figure 49. The two-day story of the 1947 Roswell ET and ETV recovery. The officer is Major Jesse Marcel, who saw the real wreckage but had to pose with the fake wreckage.

Two of the people who were hired to lead or to assist with the ATIC investigation later wrote books about it. One was Edward J. Ruppelt, a Second World War Air Force bombardier, trained as an engineer, who was recalled to service in January 1951 during the Korean War and assigned to the ATIC as an intelligence officer. The Project Grudge records were in disarray and Ruppelt was asked to take over the project. When Grudge was renamed Blue Book in 1951, he continued to direct Blue Book until September 1953, when he was released from service and returned to civilian life (Figure 50), (71,72).

The other writer was J. Allen Hynek, an astronomy professor at Ohio State University in Columbus, Ohio, which is close to Wright-Patterson AFB in Dayton. He was hired as an ATIC consultant in 1948. His job was to find astronomical explanations for UFO observations, which he did until Project Blue Book was closed in 1969. Ruppelt and Hynek were both honest people doing an honest days' work under conditions that, regardless of what they thought, guaranteed that the "extraterrestrial hypothesis:" the possibility that some at least of what were then called UFOs might be extraterrestrial vehicles, would never be confirmed by the Air Force (Figure 50).

Ruppelt had access to all of the older ATIC files from projects Sign and Grudge. While looking through the old documents he found a 1948 Project Sign intelligence estimate that summarized the project's conclusion: some UFOs were probably extraterrestrial. That 1948 *Estimate of the Situation* went up the Air Force chain of command until it reached the Chief of Staff of the Air Force. He rejected it. ATIC sent a delegation to the Pentagon to make the case in person. The Chief of Staff again rejected it. All but a few copies were destroyed and the intelligence estimate that some UFOs might be extraterrestrial was stricken from the record (71, 72).

During Hynek's longer stay as a consultant for ATIC he found that people did mistake astronomical phenomena: planets, stars, meteors, comets or other rare events, for ETVs. But, like

Ruppelt, he read many ATIC reports of maneuvering objects seen at close range by trained observers that were classed as "explained" – but with specious explanations. His book details his growing dissatisfaction with the inaccurate explanations that led to his transformation from an Air Force consultant into a pioneer ET researcher (73).

**Figure 50. Edward J. Ruppelt
and J. Allen Hynek**

The Washington, D. C. Flap

Widely publicized visual and radar sightings of ETVs peaked in the summer of 1952, and things came to a head for both Ruppelt and Hynek during the "Washington Flap" (described in the *Truth* section, p. 31). Ruppelt, head of Project Blue Book at the time, was twice called to Washington during the flap. The first time he was summoned to the Pentagon to explain to the top Air Force Brass what was going on. He didn't know, but suggested that some of the radar reports might have been weather anomalies. Ruppelt was a junior officer low on the Air Force chain of command, and he had neither the resources nor the Pentagon connections to make much of an impression on either the government officials or the media reporters who were following the Washington flap. He returned to Dayton, having neither explained the Washington sightings nor talked to the radar operators, pilots and ground observers who had seen them.

He was summoned back again because ETVs continued to fly over restricted zones like the Capitol and the Washington National Airport. This time he was called by President Truman's Air Force aide, a brigadier general, and asked what was going on. Captain Ruppelt again told the brigadier general that the radar targets might have been weather anomalies. He also attended a press conference on July 29 where Air Force major general John Samford mentioned "…in the order of twenty percent of the reports, that have come from credible observers of relatively incredible things." But another ATIC junior officer at the press conference, captain Roy James, said that the radar targets might have been caused by weather. On Wednesday, July 30, the front page of the New York Times reported that the "Air Force Debunks 'Saucers' As Just 'Natural Phenomena' " The press accepted the weather phenomenon explanation and that is how the "Washington Flap" ended (Figure 51).

Air Force Debunks 'Saucers' As Just 'Natural Phenomena'

Intelligence Chief Denies a Menace Exists —'Objects' Believed to Be Reflections, but 'Adequate' Guard Will Be Kept

By AUSTIN STEVENS
Special to THE NEW YORK TIMES.

Figure 51. New York Times headline, July 30, 1952

President Truman was dissatisfied with the confusion and uncertainty that surrounded the Washington flap. He called in the CIA. But the CIA had already made up its mind:

> "... on three of the main theories in explanation of these phenomena, - a US development, a Russian development, and space ships - the evidence either of fact or of logic is so strongly against them that they warrant at present no more than speculative consideration. ... there are many who believe in them and will continue to do so in spite of any official pronouncement which may be made ... there is a fair proportion of our population which is mentally conditioned to acceptance of the incredible. ... these sightings could be used from a psychological warfare point of view either offensively or defensively. ... the Civilian Saucer Committee in California has substantial funds, strongly influences the editorial policy of a number of newspapers and has leaders whose

connections may be questionable. Air Force is watching this organization because of its power to touch off mass hysteria and panic. ... we, from an intelligence point of view, should watch for any indication of Russian efforts to capitalize upon this present American credulity... Our air warning system will undoubtedly always depend upon a combination of radar scanning and visual observation.... yet at any given moment now, there may be a dozen official unidentified sightings plus many unofficial. At the moment of attack, how will we, on an instant basis, distinguish hardware from phantom? (74)"

The Robertson Committee

The CIA set up a committee to review the Washington flap and to recommend a course of action. Howard P. Robertson, the committee chair, was an eminent physicist who had worked closely with the US government during the Second World War. Hynek was not a member of the Robertson Committee but he was called to Washington to brief the Committee when it met in January 1953.

The Robertson Committee report stated that *public interest* in UFOs was a threat to national defense:

> ... the evidence presented on Unidentified Flying Objects shows **no indication that these phenomena indicate a need for the revision of current scientific concepts** [my emphasis; see the comment below about

twentieth-century science]. …continued emphasis on the reporting of these phenomena does … result in a threat to the orderly functioning of the protective organs of the body politic. We cite as examples the clogging of channels of communication by irrelevant reports…being led by continued false alarms to ignore real indications of hostile action, and the cultivation of a morbid national personality…hysterical behavior and harmful distrust of duly constituted authority (74).

The Robertson panel concluded (in secret) in 1953 what the CIA had already concluded (in secret) in 1952: UFOs were misperceptions, "UFO reports" were a threat to national defense and the public should be discouraged from taking an interest in or reporting UFOs (75).

Science and the Washington Flap

In *UFOs, ETs and Alien Abductions* (1) I described how academic science from the late nineteenth to the mid-twentieth century was oriented towards developing and testing theories, and unless an observation provided evidence to either support or attack a specific theory, the observation itself was of little interest. That description – and it was unfavorable -- originated with many eminent philosophers of science including William James and Thomas Kuhn (76, 77). When ETVs first made headlines in the mid-twentieth century, there was no accepted theory to explain how they worked – except the theory that ETV reports were errors of human vision or of human memory. In other words: all ETV observations were psychological mistakes. The CIA and the

Robertson Committee and Project Blue Book all believed that. Most scientists believed that then. Some scientists still believe that now.

Following the Robertson Committee recommendations, and after the Washington Flap ended, Project Blue Book continued to "investigate" ETVs and explain them as misperceptions often based on astronomical phenomena. Blue Book contracted with the Battelle Memorial Research Institute in Columbus. Ohio, to transfer data from about 4,000 Blue Book sighting files collected from 1947 to 1952 to punch cards (the data entry tool of the nineteen-fifties), and then to analyze the reports and to summarize their findings. To emphasize the prevalent attitude towards observations in that era, I quote a sentence from the introduction of the Battelle report: "It must be emphasized, again and again, that any conclusions contained in this report are based NOT on facts, but on what many observers thought and estimated the true facts to be." And their conclusion, expressed in the last sentence of their 1955 report : "Therefore, it is considered to be highly improbable that any of the reports of unidentified aerial objects examined in this study represent observations of technological developments outside the range of present-day scientific knowledge" (78).

Note the similarity between the 1955 Battelle text and the 1952 Robertson Committee text cited above. Both the Robertson Committee and the Battelle Institute had only one theory to explain ETV observations: ETV observations were psychological errors.

Donald H. Menzel and Philip J. Klass: Professional Liars

Motivated by prevailing scientific opinion and encouraged by contacts with members of the Robertson Panel, one distinguished scientist and one scientific journalist became "debunkers" and began to write books and articles to re-educate the ignorant and self-deluded public who kept on seeing and reporting ETVs. The two outspoken opponents of the extraterrestrial hypothesis were the astronomer Donald H. Menzel (1901-1976) and the science writer Philip J. Klass (1919 – 2005) (Figure 52). Menzel was a productive scientist whose contributions to astronomy were widely known and respected. Klass, trained as an electrical engineer, was a journalist who wrote regularly for and served as editor of *Aviation Week and Space Technology*, the leading American magazine reporting on the aerospace industry. They both vehemently upheld the psychological error theory. Between them, from 1953 to 1997, they wrote ten books that "debunked" the extraterrestrial hypothesis (79). Menzel, Klass and Blue Book were disinformation tools in the campaign against the UFO evidence and against the private groups that investigated it.

Figure 52. Donald H. Menzel and Philip J. Klass

Seth Shostak and Sara Seager:
Astronomers with limited vision

Contemporary astronomers Seth Shostak and Sara Seager are competent professionals who can see clearly through the giant eyeglasses of their trade (optical or radio telescopes) but cannot see what is in front of their noses. As I explained in the introduction, they also belong to the class of scientific people who dismiss what they do not understand (pp 1-2). Shostak calls ETVs (UFOs) "an enormously widespread and durable fantasy," and wrote that "An abundance of equivocal sightings fails to add to a compelling case for visitors." (80). He devotes 42 pages of a book about his career in SETI (using radio-telescopes to search for extra-terrestrial intelligence) to opine about ETVs. His conclusion – restated by me in many fewer words than he actually used – it can't be; therefore it isn't. Sara Seager, another SETI astronomer whose research tools are radio-telescopes and satellite sensors, answered questions about ETs in a newspaper interview:

"Q: So do you think we've had visitors from other planets here? A: I don't actually. People love believing that; I'm not quite sure why so many people believe ... But no, the distances are so vast and that's not why (I don't believe), but because there's no hard evidence. None of the arguments are very helpful or offer any kind of proof" (81).

You have read page after page of eyewitness and instrumental evidence. Why don't these near-sighted astronomers accept the evidence? Because the ETV and ET evidence does not come from their scientific tools, which are optical or radio telescopes, it is not within their professional sphere of competence and they feel free to ignore it. Because there is no accepted scientific theory that explains how ETVs might work, the evidence is uninteresting to them. They have probably never taken the trouble to read the evidence, let alone to study it. These near-sighted astronomers believed what they said. That is no excuse for offering a "professional" opinion concerning something about which you are simply uninformed. (Figure 53).

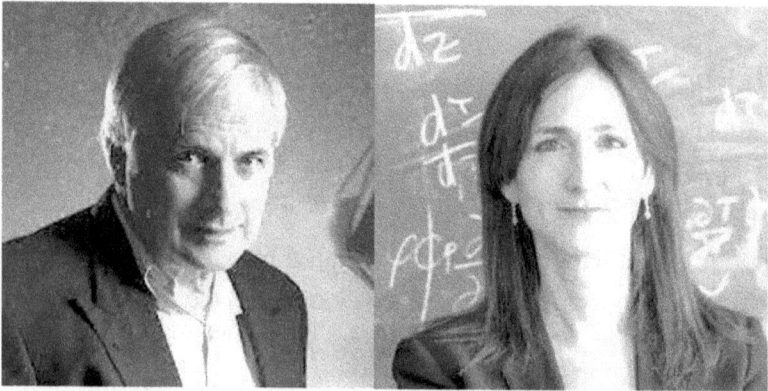

Figure 53. Seth Shostak (SETI Institute), Sara Seager (Justin Knight)

8. The Good Side

Clear-sighted Astronomer: Avi Loeb

Avi Loeb (Figure 54) is the Harvard astronomer who suggested that a celestial object that was seen near the sun through a Hawaiian telescope in 2017, and that he called "Oumuamua," might be a scout craft from an ET civilization. He is, in effect, telling fellow-astronomers Seth Shostak and Sara Seager to wake up to the possibility that ETs might be visiting; that ETs might be interested in our solar system and just might be interested in our civilization. That puts him on the Good Side. He agrees with Samuel Johnson that: "Human experience, which is constantly contradicting theory, is the great test of truth."

Loeb supports the need to study the possibility that we are not alone. He praised the Pentagon report on Unidentified Aerial Phenomena (UAP, formerly UFOs) that was released in June 2021 (3), writing that

> Former top-level government officials who had access to the classified UAP data, including former U.S. President Barak Obama, former Intelligence Director John Ratcliffe, former CIA Director James Woolsey and former Senator Harry Reid stated that they believe UAP is a serious matter, worthy of scientific inquiry. If so, how can scientists – who are supposed to be open-minded, avoid

the inquiry? (82).

They can't.

Figure 54. Astronomer Avi Loeb

Private ET Study Groups

Private groups have studied ETs and ETVs for almost as long as governments have lied about ETs and ETVs. A few of the currently active US groups, borrowed in part from a more comprehensive Wikipedia list, are named below (Figure 55). This is not a comprehensive history of North American or world-wide ET study groups, but Wikipedia's list, together with the many links in the Wikipedia article to other scholarly sources, will satisfy the curious reader (83).

Some Active American ETV Study organizations
Center for the Study of Extraterrestrial Intelligence
Center for UFO Studies
International UFO Congress
Mutual UFO Network
National Aviation Reporting Center on Anomalous Phenomena
National UFO Reporting Center

Figure 55. Some currently active US ET and ETV private research groups

Here is a brief historical review of some of the better-known private ET and ETV research organizations. Their interactions with the public were all undertaken in opposition to the US government's effort to tell a different story based on lies.

The Aerial Phenomena Research Organization (APRO) was founded by Jim and Coral Lorenzen in 1952. Coral had seen an ETV in 1934 when she was a young girl living in Wisconsin. She saw another ETV, in Arizona, on June 10, 1947, at about the same time (June 24, 1947) that Kenneth Arnold saw his ETV. Coral Lorenzen and her husband Jim knew that Coral's sighting and the Arnold sighting were part of a larger picture. They started a group to record and investigate ETV sightings, and they began to publish the *A.P.R.O. Bulletin* in 1952. APRO moved with the Lorenzen family as jobs took them back to Wisconsin and then again to the southwest, first to New Mexico and then to Arizona.

APRO attracted members who added technical competence to journalistic integrity in reporting both ET and ETV cases to the public. Publication of the *A.P.R.O. Bulletin* continued until September 1987. After APRO closed, many of its records were transferred to what is now the Mutual UFO Network (MUFON), active since the mid nineteen-sixties.

The National Investigations Committee on Aerial Phenomena (NICAP) was founded in 1956 by Thomas Townsend Brown, who you have already met: he invented the Gravitator, the first in a long line of patents and experiments that are leading us towards mastering interstellar travel (see p. 50 and Figure 20). NICAP was based in Washington, D.C. and its management included people who were either well-connected to the U.S. government or who had been in government themselves.

One board member was Roscoe H. Hillenkoetter, a retired Navy admiral who had served in both World Wars and had worked his way up through the intelligence community to become the first director of the CIA. Another was Joseph Bryan III, who served in the Navy during the Second World War and, afterwards, with the CIA. Joseph Bryan's son, the late C.D.B. Bryan, also took an interest in ETVs. He wrote *Close Encounters of the Fourth Kind* (Knopf, 1995), a report about the Abduction Study Conference (36).

NICAP was antagonistic towards the US government because many of its directors had first or second-hand knowledge of the kind of obfuscation that happened during the Washington Flap, even though the CIA memo and the Robertson committee report, described earlier, were still secret at the time. NICAP published a newsletter, *UFO Investigator*, from

1956 to 1980. It also published *The UFO Evidence*, edited by Richard H. Hall, in 1964 and *The UFO Evidence: a Thirty-year Report,* also edited by Hall, in 2001. Both are valuable compendia of carefully analyzed case studies. I joined NICAP in 1966 and helped NICAP with a 1968 study of humanoid sightings (31).

From 1975 through 1981 I belonged to UFO-Québec, a research group that published French-language reports, including some of mine, in 25 quarterly issues of its *Informations et Recherches*. I also became a member of the Center for UFO Studies (CUFOS, Now the J. Allen Hynek Center for UFO Studies, www.cufos.org), and MUFON (The Mutual UFO Network, www.mufon.com). MUFON has an international presence and I currently work with both local and international MUFON members to study and evaluate reports involving contact with ETs.

J. Allen Hynek made a cameo appearance before the Robertson Committee, and then went back to his day jobs – as an astronomy professor in Columbus, Ohio and as a part-time astronomical consultant to Project Blue Book. He continued to look for astronomical explanations for cases he was asked to review by Blue Book, but became increasingly upset by Blue Book's failure to thoroughly analyze cases that did not have astronomical explanations. (This followed naturally from Blue Book's role after the secret Robertson Committee report, as described above). Hynek described his growing disenchantment with Blue Book and with the Condon Committee (to be discussed shortly) in *The UFO Experience: A Scientific Inquiry* (73), which he published in 1972. After he resigned as a consultant for Bule Book, he founded the Center for UFO Studies, listed in Figure 55.

Hynek wrote the article titled "Unidentified Flying Objects" for the 1969 edition of Encyclopedia Britannica. It included the table reproduced below, accompanied by this text::

"The steady rise in reports from 1963 to 1966-67 is worthy of note. Indeed, one might base a call for serious scientific attention to the UFO problem on the reports of 1965 and 1966 alone" (Figure 56).

ETV Sightings from 1947 through 1966
(Encyclopedia Britannica, 1969, Vol. 22, p 500)

Reported Sightings, 1947-1967

Year	Number	Year	Number
1947	79	1958	599
1948	143	1959	364
1949	186	1960	515
1950	169	1961	488
1951	121	1962	474
1952	1501	1963	399
1953	425	1964	526
1954	429	1965	887
1955	404	1966	1060
1956	778	1967	1000
1957	1178*		

* Results incomplete.

Figure 56. From Volume 22 p. 500 of the Encyclopedia Britannica, 1969 (J. Allen Hynek, author)

Hynek did more than write an encyclopedia article; he also wrote to the head of the Foreign Technology Division at Wright-Patterson AFB (the successor to ATIC) and suggested that a scientific panel should be re-established to review the accumulated ETV data.

Other scientists had come to the same conclusion. James E. McDonald was an astronomer at the University of Arizona whose research specialty was cloud physics. McDonald and two meteorologist colleagues saw a UFO on January 10, 1954. They found neither a commonplace nor a scientific explanation for what they saw. McDonald was interested enough to start learning about the groups and the people studying the UFO phenomenon. As sightings increased through the 'sixties he began to ask his professional colleagues, including some in government, to take the sightings seriously.

McDonald had a legitimate professional interest in the reports collected by Project Blue Book because they might contain information about cloud formations, one of his research interests. He applied for and got a research grant to visit Wright-Patterson field and to search the Blue Book files. While looking through the files he found the Robertson Committee report. He learned from the report that Blue Book was a cover-up, and he learned from the Blue Book files that UFO reports were reliable. This encouraged him, together with people like Hynek, to agitate through political channels for the government to go beyond Project Blue Book and make a better effort to understand the phenomenon. Together with the effect of growing public interest on Congress, the scientists' agitation succeeded. In 1966, the Air Force announced that it was going to establish and fund an independent scientific committee to investigate the UFO phenomenon. And in 1968, the US House of Representatives held a hearing on Unidentified Flying Objects.

The Dark Side intrudes: The Condon Committee

The UFO sighting increase in the mid nineteen-sixties shown in Figure 56 was widely and often seriously reported in the press and on TV, despite Blue Book's efforts to trivialize the observations. That increase also caught my attention as a student of human visual perception and memory, and, as I explained earlier, led me to take a professional interest in what people said they were seeing that the government said they were not seeing.

The Air force asked many universities and laboratories to carry out the new UFO study and most of them refused. The Air Force finally contacted Edward U. Condon of the University of Colorado's Physics Department and asked him to head the study. Condon had worked on the Manhattan Project (the US atomic bomb project) during the war, he headed the National Bureau of Standards after the war, and he had published many physics books and articles. Condon and the University of Colorado were reluctant to take on the project until Robert J. Low, assistant dean of the university's graduate school, wrote an internal memo to explain how the university could accept the project without losing its credibility:

> The trick would be, I think, to describe the project so that, to the public, it would appear a totally objective study but, to the scientific

community, would present the image of a group of nonbelievers trying their best to be objective but having an almost zero expectation of finding a saucer. One way to do this would be to stress investigation, not of the physical phenomena, but rather of the people who do the observing—the psychology and sociology of persons and groups who report seeing UFOs .... if we set up the thing right . ... we could carry the job off to our benefit (84).

The University accepted the project and Condon chaired the committee. He left its day-to-day operations to Robert Low. In late 1967 Condon said to a scientific audience, "My attitude right now is that there's nothing to it . . . but I'm not supposed to reach a conclusion for another year".

The Condon Committee Report was published in 1969 (12). The paperback version has 965 closely printed pages and is nearly half a million words long. Much of it consists of appendices and chapters describing materials and methods that have nothing to do with the cases studied. The important parts of the report are the first chapter, written by Condon, and the case studies that were prepared and written by members of the committee staff (the Condon Committee's McMinnville photo case was described in the *Truth* section, p. 18).

The Condon Committee followed a path that J. Allen Hynek and James E. McDonald had seen and complained about in Blue Book. Researchers analyzed the information provided from field investigators and witness reports. They found no conventional explanations for many cases and classified them as "unknowns," but the "unknown" cases were ignored in the report's "Conclusions and Recommendations:"

> "Careful consideration of the record as it is available to us leads us to conclude that further extensive study of UFOs probably cannot be justified in the expectation that science will be advanced thereby If they [scientists] agree with our conclusions, they will turn their valuable attention and talents elsewhere. We believe that the rigorous study of the beliefs— unsupported by valid evidence—held by individuals and even by some groups might prove of scientific value to the social and behavioral sciences . . The question remains of what, if anything, the federal government should do about the UFO reports it receives from the general public. We are inclined to think that nothing should be done with them. The subject of UFOs has been widely misrepresented to the public by a small number of individuals who have given sensationalized presentations in writings and public lectures . . whatever effect there has been bad (12, p. 1)."

The Robertson Committee recommended lying about UFOs, so the Condon Committee lied about UFOs. The Condon Committee report dismissed the scientific value of UFO reports and discouraged scientists from studying them. The report castigated irresponsible people for misrepresenting the evidence, and opined that reading UFO books and magazines was bad for children. Condon unburdened the Air Force, which had already told him exactly what it wanted, from pretending to study UFOs. Project Blue Book was closed shortly afterwards in 1970.

The 1968 Congressional UFO Symposium

While the Condon Committee was still at work, and before it had released its report, the US House of Representatives held a UFO hearing of its own. It was called *The Symposium on Unidentified Flying Objects: Hearings before the Science and Astronautics Committee of the U. S. House of Representatives*. It was held on July 29, 1968 (85). The symposium heard first from J. Allen Hynek and then from James E. McDonald. A third speaker was Carl Sagan, a well-known astronomer. There were three other speakers, all scientists and engineers. The symposium also published written submissions from Donald Menzel, engineering physicist Stanton Friedman (who wrote about the Barney and Betty Hill case) and four other scientists.

Hynek and McDonald testified that the human observations and instrumental recordings were reliable, that the evidence required serious consideration, and that one possible explanation was that some of what people were reporting as UFOs were ETVs. They strongly urged continuing government-supported research to investigate the ETV hypothesis. Their written statements made the same recommendations.

Carl Sagan started by saying that it was unlikely that there was other intelligent life in our solar system and that it was even less likely that ETs from other solar systems could travel here because the speed of light was an upper speed limit for travel and the nearest other star was 4 ½ light-years away. He then explained that people want to believe in gods and extraterrestrial beings; that the ET belief satisfied the same human needs that were satisfied by religion, that people did not subject their religious beliefs to the same high standard of proof as are used for scientific proof, and that the evidence was insufficient to justify further interest in UFOs by the government. He finished by making a pitch for funding his own fields of interest: "..if Congress is interested in a pursuit of the question of extraterrestrial life, I believe it would be much better advised to support the biology, the Mariner, and Voyager programs of NASA, and the radio astronomy programs of the National Science Foundation, then to pour very much money into this study of UFOs."

All opinions were made public at the Symposium – but it made no difference to the US government, because the Condon Committee report, released six months after the congressional symposium, led to the closing of Project Blue Book, the last publicly acknowledged US organization with an interest in ETVs.

The American Association for the Advancement of Science (AAAS) Symposium

In December 1969, about a year after the Condon Committee Report was released, the American Association for the Advancement of Science (AAAS), a private science organization, held a symposium on unidentified flying objects. Edward Condon and Donald Menzel tried unsuccessfully to persuade the AAAS to cancel the event. Condon refused to participate, but Menzel showed up to oppose the "nuts" and "believers," his terms for scientists willing to consider the extraterrestrial hypothesis. The symposium papers were edited by Carl Sagan and Thornton Page and published in 1972 as *UFOs: A Scientific Debate* (86).

William K. Hartmann, an astronomer and photo expert who had analyzed the McMinnville, OR case study for the Condon Committee report, wrote about the extreme difficulty of determining absolutely that a photograph is genuine. Philip Morrison, an MIT physicist, commented on the difficulty of establishing an unequivocal chain of evidence about a novel event. Some contributors described UFO witnesses as inherently irrational, explaining that "primary-process thinking"— mental processes related to fundamental motivations—makes it difficult for witnesses to report their experiences objectively. Others suggested that UFOs could be integrated into a college curriculum to attract interest to basic science. Donald Menzel contributed "UFOs—The Modern Myth," which covered cases that he claimed had atmospheric or optical explanations. J. Allen Hynek summarized his twenty-one years of studying UFO reports and stressed the consistency and uniqueness of those that remained unexplained after thorough investigation. James E. McDonald's (Figure 57) 62 – page paper was titled: "Science in Default: Twenty-Two Years of Inadequate UFO Investigations." He directly criticized Condon Committee dismissals of four of the cases that they reviewed, including the RB-47 case that was presented in the *Truth* section of this book (p. 33). McDonald concluded: "There are enough significant unexplainable UFO reports within the Condon Report alone to document the need for a greatly increased level of scientific study of UFOs."

McDonald's professional life depended on government grants and contracts. Debunker Philip Klass approached McDonald's government research sponsors and urged them to cancel his research grants. When a congressional committee investigating supersonic transports called on McDonald for testimony about the effect of supersonic flight on the atmosphere, a congressman ridiculed. McDonald's UFO interest in order to undermine his credibility. As McDonald's UFO commitment increased, his home life fell apart. His wife left him. A failed suicide attempt left him blind. He killed himself in 1971.

Figure 57. Astronomer James E. McDonald, hounded to death by UFO debunkers

Science and ETVs after 1969

Science, the general science journal of the American Association for the Advancement of Science (AAAS), published little on UFOs from the 1947 Arnold sighting to the present. In 1966 they reluctantly published a shortened version of J. Allen Hynek's critique of the Condon Committee Report. In 1967 they published a paper explaining that UFOs could not be extraterrestrial because an explanation could not be found for how UFOs could get from space to Earth. They also published three UFO-related news articles (two about the Condon Committee, one about a private 1998 UFO symposium), one book review, and thirteen letters, one of which was from my late colleague and collaborator, Stuart Appelle, who wrote that science is legitimized by its methodology, not by the subject matter it investigates (87) .

The only UFO-related empirical research paper *Science* ever published was a sociological article called "Status Inconsistency Theory and Flying Saucer Sightings". Author Donald Warren suspected that people whose incomes didn't match their education levels—and who suffered status-related social anxiety thereby—sought to reduce their anxiety by getting attention, social recognition, and personal satisfaction through reporting a UFO sighting. Warren tested this idea against Gallup poll demographic data for people who had made UFO reports. He was wrong about the entire sample containing African-Americans, Hispanics, and white Americans. But he was right about white UFO reporters: proportionally more UFO reports came from status-inconsistent than from status-consistent white people (88). The report did not address the nature or credibility of the reports. It simply confirmed that if you were a white person in 1971, the more consistent your social status, the less likely you were to report having seen a UFO. The establishment gatekeepers for a prestigious scientific journal took a characteristic approach to UFO evidence: don't ask *what* are these

Unknowns? but ask *why* are these Unknowns? According to Warren, because the reporters were social status–seekers.

A Summary of The Lies

This statement used to appear on the US Department of Defense's website (89):

- No UFO reported, investigated and evaluated by the Air Force was ever an indication of threat to our national security

- There has been no evidence submitted to or discovered by the Air Force that sightings categorized as "unidentified" represented technological developments or principles beyond the range of modern scientific knowledge; and

- There was no evidence indicating that sightings categorized as "unidentified" were extraterrestrial vehicles.

The page is no longer available. Neither are its lies.

ETs

The future depends on what you do today.

Mahatma Gandhi

ETs challenge human civilization. *ETs Imagined and Real* separates our artistic fantasies from the ET reality. Recognizing and then meeting *The ET Challenge* is how we thrive as a civilization and survive as a species. If we ignores the ET challenge, humanity relinquishes control over its future: your future, my future and our children's future.

10. ETs: Imagined and Real

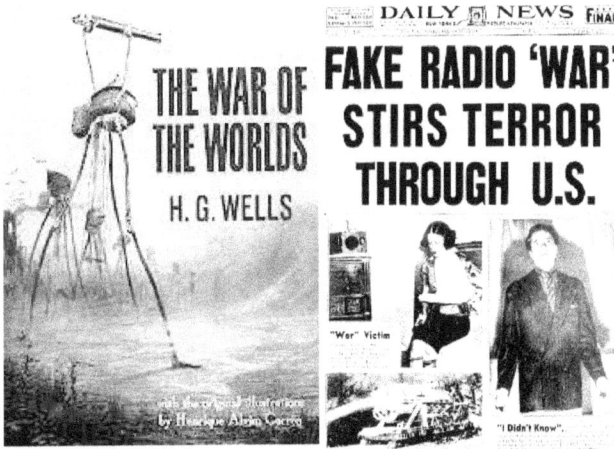

Figure 58. H. G. Wells' novel, The War of The Worlds , and headlines that followed the "live" broadcast" as if it were happening in New Jersey.

H.G. Wells' The War of the Worlds, an 1897 novel about Martians invading London (Figure 58), (90) came alive one evening in 1938 when a radio adaptation by Orson Welles interrupted a music program and was presented as "breaking news." The "news" was that Martian rockets had landed in New Jersey. Budd Hopkins, then a child of seven and later an artist and abduction investigator, listened to the broadcast at his home in Wheeling. West Virginia. One of his father's panicked employees called to say that he was heading for the hills to escape the Martians (32, pps. 23 – 27).

We were immersed in ET fiction later in the twentieth century (Figure 59). Captain Kirk of *Star Trek* boldly went where no man has gone before. *Star Wars* ETs included Chewbacca and Jabba the Hutt. Who could forget *ET, the Extra-Terrestrial* or the Vulcan-human *Star Trek* hybrid, Spock? Movies and TV made ETs a backdrop for late twentieth-century life, and our children, nephews and nieces grew up with them.

ETs routinely walk through contemporary literature in cartoons, one of which (Figure 60, left) is a visual takeoff on the Welles' radio adaptation of *The War of the Worlds*.

Figure 59. Fictional ETs: ET and Spock, the Vulcan-Human hybrid

"Until we hear different, it's Jersey's problem."

"Who do we talk to about buying your planet?"

Figure 60. ETs in *New Yorker* cartoons.

From Imagination to Reality

Astronomers think that there may be as many as 30 earth-like planets orbiting stars that are less than fifty light—years away from earth, and that are within the "habitable zone" of each star, meaning that the planet has a surface temperature similar to earth's (91). Scientists and the rest of us accept that these earth-like planets exist. If the ETs that you met in Chapters 4 through 6 thrive under conditions like ours, then one or more of these earth-like planets could be their home.

There are many remembered versions of what ETs look like. Given that there are many possible planets of origin for our visitors, and many humans who have seen ETs either as abductees or during a close encounter on earth, it is not surprising that there may be real differences among the features of the ETs that have been recalled. This ET

stereotype is based on evidence that is consistent across many witnesses, and you will recognize it from the Roswell story the short grey with the oval eyes, pointed chin and small facial features (Figure 25, p. 66 and Figure 61, below).

The pointed chin, elongated eyes, small or nonexistent ears

Figure 61. ET face remembered and drawn by an abductee. Hopkins (44), Figure 19

and nose, and thin mouth are commonplaces of ET appearance. Other people report ETs who are taller (Figure 62), with insect-like characteristics of body and or face (Figure 29, p. 75), or of a reptilian appearance of body and skin (Figure 62). Human-appearing ETs have also been reported.

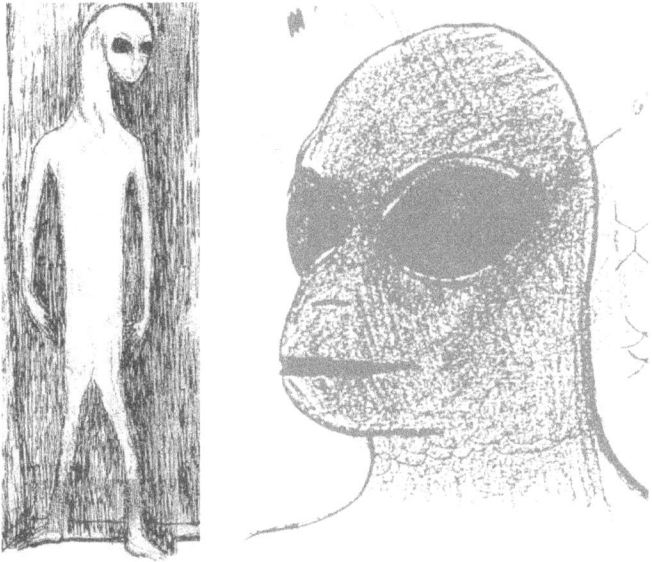

Figure 62: "Tall grey" and "reptilian" ETs

But we still don't know how they get here. The summary of possible ETV technology reported in Chapter 3: *How do ETVs work?* does not tell us *how* ETVs work, it tells us how they *might* work. Chapter 3 does show us that we are working on the problem. There is the beginning of a scientific understanding of how something as large and heavy as an ETV could be made to move through space fast enough to get ETs from stars as far as fifty light-years away to earth in a "reasonable time."

What is a "reasonable time"? Fifty light-years (the distance light, travelling at about 300,000 km/sec, travels in 50 years), is about 500 trillion kilometers. But time – time itself – slows as the traveler moves faster; and the closer the traveler comes to moving at the speed of light, the slower the traveler's time passes. A traveler moving at the speed of light would be suspended in a timeless world. This follows from Einstein's theory of special relativity. Paul R. Hill (1909-1990) was an aeronautical engineer who designed the fuselage of the World War 2 American P-47 fighter-bomber. After having seen several UFOs, he wrote *Unconventional Flying Objects* (92), published posthumously in 1995. In Chapter 18, he wrote:

> .. there is an important distinction to be made between the time experienced by the space traveler and the time which passes meanwhile on the home planet and on the planetary destination. The tremendous acceleration, speed and energy capabilities displayed by UFOs make them well suited to capitalize on this distinction by the attainment of reduced on board times realizable by approaching the speed of light (92, p.278).

Hill recognized that space travel will be complicated. A friend who is your age could step into an interstellar ship, fly away, and come back, three years older, into a world where you and his other stay-at-home friends had been long in the grave. No one said that joining the universe will be easy, or that, after we join, our emotional and intellectual worlds will be the same.

Science Fiction, Indifferent Science and the ET Challenge

Much of what we know about ETs and ETVs has been reported here. This public ET and ETV database has been created entirely by people and groups who are outside the "scientific establishment." They are not among those who work for or belong to the associations, publishing houses, universities and research labs that generate the information that informs the public about the world as it is understood by institutional science. I explained why this was true in my earlier book, *UFOs, ETs and Alien Abductions* (1). The consensus-driven engine of public research funding limits professional scientists' financial support to studying only those things that most of their peers think are worth studying. It surprises me, given the recent evidence described in Chapter 1 (*ETVs -Yesterday*), that the ET and ETV observational and instrumental database is still not getting the attention it deserves from the scientific establishment. But as I write, professional scientific attention is beginning to shift because ETV evidence is being taken seriously by a few more academics and politicians, reliable information is now available through respected public media, and private organizations are beginning to give our ET and ETV problem more attention.

The science-fiction world of ET literature and drama, combined with rejection of ET and ETVs by the scientific establishment, has made it easy to push the anxiety-provoking facts about ETs and ETVs out of mind, either as pure fiction or as "not scientific." But we know that ETs and ETVs are real and cannot be ignored, and we know that we have to face and overcome the ET challenge together.

11. The ET Challenge

ETs control us telepathically when they are near us. They use telepathy to abduct (kidnap) and examine humans aboard ETVs. ETs extract sperm from men and inseminate women with hybrid ET-human sperm. The women conceive hybrid fetuses. Women who are pregnant with hybrid fetuses are then re-abducted and the developing fetuses are extracted and brought to term aboard the ETV. Some mature ET-human hybrids are returned to earth and become part of human society.

There are two technical parts and one political part to the ET challenge. The technical parts are the telepathy challenge and the technology challenge. The political challenge is for humans to learn to act together as a species to survive ET interference. The telepathy challenge is being studied by private citizens and the technology challenge is certainly being studied by governments. There may be partial solutions to the telepathy challenge and to the technology challenge, but the political challenge has been neither widely considered nor solved.

The Telepathy Challenge

We have to learn how to stop ETs from forcing us to do what they want us to do. They want us to let them meddle with our reproductive systems to create ET-human hybrids, some of whom may look enough like us to take up residence among us, and who are equipped with persuasive telepathic abilities (43). The tools they use for doing this have been discussed earlier (pps 127 – 133).

The Technology Challenge

I explained earlier that we may be learning how to do what ETs already know how to do: reduce gravity and inertia and accelerate close to the speed of light. Research that may show an interaction between emg energy and mass is described on pages 51 – 59. Inventions suggesting that gravity can be modified and energy can be obtained by the controlled manipulation of emg radiation are described on pages 51 – 56. The research area is active and progress is being made, as suggested by both the experimental evidence and the published patents. Patent documents and eyewitness accounts suggest that aspects of ETV technology may already have been incorporated into vehicles that the US government or its contractors know how to make (pages 56-59).

The Political Challenge

The Universal Declaration of Human Rights, proclaimed in 1948 by the United Nations General Assembly (94), includes these articles:

- Article 3. Everyone has the right to life, liberty and the security of person…

- Article 12. No one shall be subjected to arbitrary interference with his privacy, family, home or correspondence, nor to attacks upon his honor and reputation.

Everyone has the right to the protection of the law against such interference or attacks…

Articles 3 and 12 tell us not to kidnap people against their will and not to interfere with their privacy, family and home. Governments try with varying degrees of sincerity and success to protect those rights from violation by governments themselves or by other people, but no government appears ready to protect those rights against violation by the ETs who routinely kidnap people and arbitrarily interfere with their privacy and their happiness.

Government inaction about ETs may change. More than thirty percent of Americans thought in 2020 that ETVs (still called UFOs by opinion pollsters) are real, and ten percent of them said that they have seen an 'alien spaceship' In a 2019 Gallup poll, sixty-eight percent of Americans thought that the U.S. government knows more about UFOs than it is telling us – I agree (94). The US government has stopped lying about ETVs, which is long overdue. But it is not telling us all that it knows about ETs and ETVs – at least all of what some part of the U.S. government knows.

While thirty percent of Americans think that ETVs are real, about two percent may have experienced an ET abduction (page 70). Although the thirty percent who accept the reality of ETVs may not yet be ready to accept that two percent have been abducted by ETs, the purpose of this book has been to persuade the thirty percent as well as the rest of us that ETs are real and that they represent a threat to the stability and independence of human civilization.

The Existential Challenge

Our family and local traditions, our historical cultures and our institutions of government provide us with the memories and expectations that allow us to feel that we are managing, however imperfectly, our own lives, our own countries and our own planet in concert with the 8 billion or so other people who live on earth. In other words, we act as if we control our own destiny.

A dispassionate observer might question how well we are doing as a species, and it is a fair question. Our political institutions are no guarantee against thermonuclear war, our exploitation of planetary resources has created environmental problems without a consensus to resolve them and our imperfect societies do not protect the most vulnerable humans against lives that can be "nasty, brutish and short" (95). The hopeful among us believe that humans can solve these problems by improving our political institutions, advancing our understanding of nature and evolving a planet-wide culture that is compassionate towards everyone and respectful of earth.

Those aspirations will be superfluous if we no longer manage the earth by ourselves. ETs possess the technology to master both us and earth if they choose to do so. Their ETVs are technologically more powerful than our most advanced machines and ET telepathy can control human behavior more effectively than a resolution of the UN General Assembly, a law passed by your national legislature or a bylaw of your city council. Whether ETs will continue to interfere with us and our planet is not clear to me. That that they have the tools to do so – advanced ETV technology and telepathic control – is clear to me.

In order to avoid becoming subordinate creatures on our own planet, like the family pets who we bond with emotionally but not intellectually, or the zoo animals that we confine and feed so we can enjoy looking at them, we must negotiate our status with ETs to guarantee that we are treated as equals on our own planet and in the universe. ETs may just be trying to be helpful, and they may be willing to treat us as equals, but we do not know that and cannot take it for granted: we have to negotiate our status as equals with the ETs.

How can we succeed?

Now that the Pentagon has stopped lying about ETV sightings, ETV observations are finally being evaluated on their merits. This means that ETVs and ETs will become part of consensus reality within the next few years. There are signs that this is already happening (2). Recognition of ET reality is the first step towards meeting the ET challenge.

We need to recognize and to acknowledge that ETs intrude into our life on earth and we need to persuade governments to develop and deploy the tools needed to control that intrusion. Those tools will include sensors and machines to monitor and control ETV use of our airspace. They will include material and techniques that prevent telepathic and physical control of humans by ETs. They will include sitting down with and negotiating with ETs.

The most challenging thing that ETs do to us is to produce ET-human hybrids, because the telepathically-capable hybrids have the capacity to interfere with the future of our species. This is not the fault of the hybrids but of the ET program that produces them. In the interest of preserving our species, the hybrid problem has to be confronted, controlled and resolved. This can happen only after governments, equipped with technology that matches ETV performance and that protects against ET telepathy, are able to negotiate our status as equal rather than subordinate partners in the universe into which we have stumbled.

How might we fail?

The credibility problem

Almost everything that is publicly known about ETVs and ETs has been learned by people and private groups who started to collect and distribute information about them in the mid-twentieth century. The US government started to conceal and debunk information about ETVs in 1947, and only stopped the debunking in 2017. The people who concealed ETV information knew about ETs. So did the US intelligence agencies whose responsibility, then and now, is to collect the facts needed to defend the nation. Private groups have learned a lot about ETVs and ETs, but what we know through private groups is certainly less that what some governments already know using information obtained from their powerful sensors and their other sources of intelligence. ETVs and ETs are not the fabrications of people who conjure up illusions without facts. ETVs and ETs are real, and they are better known to at least some governments than they are to us (96).

We might fail to meet the ET challenge if some of us continue to use "enthusiast" or "conspiracy theorist" as denigrating dismissals of other people's interest in and concern about the

ETV and ET evidence. The "enthusiasts" who count are the hundreds of people who have studied ETVs and ETs over the last seventy-five years, and much of what they have learned is summarized in this book. The "conspiracy" that matters was the US government's seventy years of lies about ETVs and ETs that started in 1947 and stopped, fortunately for us, in 2017.

We routinely respond with "cognitive dissonance" to a situation that has severe consequences, an unclear resolution and a plausible but specious reason to ignore it (68, p. 77). The cognitive dissonance reaction is to accept the specious explanation and to avoid the real problem. Cognitive dissonance describes how some US government agencies and their scientific supporters responded to the ETV evidence during the 1950s (pps 153 - 156), and cognitive dissonance explains why scientific skeptics like Seth Shostak and Sara Segar still ignore and dismiss the ETV and ET evidence. "Enthusiast" and "conspiracy theorist" are specious dismissals that ignorant professionals still use to avoid confronting the arguments presented by the people who know the facts about ETVs and ETs.

In *Collapse,* Jared Diamond explained how societies come to fail (97). One way is through psychological denial (aka cognitive dissonance): "If something that you perceive arouses in you a painful emotion, you may subconsciously suppress or deny your perception in order to avoid the unbearable pain, even though the practical results of ignoring your perception may prove ultimately disastrous." George Orwell put it clearly when he wrote in 1944 that "People can foresee the future only when it coincides with their own wishes, and the most grossly obvious facts can be ignored when they are unwelcome (98)".

Diamond also blames creeping normalcy: "…slow trends concealed within noisy fluctuations. If the economy, schools, traffic congestion, or anything else is deteriorating only slowly, it's difficult to recognize that each successive year is on the average slightly worse than the year before, so one's baseline standard for what constitutes "normalcy" shifts gradually and imperceptibly." Creeping normalcy describes how ETs have been hybridizing humans. If it is not stopped we may suffer the transformation or elimination of *homo sapiens* by ETs through a process that we do not control. If interaction with ETs is slowly changing our species, then that interaction and those changes should either be stopped, or continued with the consent and under the control of all of us. Human governments have to negotiate the role that ETs will play in our future. But before governments can act, they must develop the defensive tools we need, and then we must make them act.

A Long and Interesting Future

The goal of *Truth, Lies and ETs* has been to describe ETs' interactions with us and to review our collective response to the ETs. Although evidence suggests that we may be learning how to control gravity and "go interstellar," we are not there yet. In the meantime, we need to find a *modus vivendi* with ETs if we are to survive as a species among the technologically advanced civilizations in the universe.

Finding that *modus vivendi* is a job too big for one book to explain or for one person to achieve. It must be a collective enterprise. Some collective institutions represent at least nominally all of humanity. The United Nations was founded to reduce or eliminate conflicts among its 193 member countries. As the largest organization that can claim to represent humanity, it is an obvious place to start. Multinational governments like the European Union, which are more flexible and faster to respond than the UN, could also help. Private foundations and other non-governmental organizations (NGOs) with international scope and the goal of improving humanity should also take an interest because their reach is broad, their structure is more flexible and their concerns transcend local politics and the disputes among nations.

What we need to achieve is the independence of humans and human civilization from control, technologically, biologically, and telepathically, by any or all of the ETs who have come to observe, study, and sometimes interfere with us. If we do not come to an agreement with ETs about our own security on our own planet, we may be the last act in the drama of *homo sapiens* that has played out on earth for about three hundred and fifty thousand years. It will take awareness and action by all of us to guarantee that the future of our species is a long and interesting one.

Afterword

"Earth is in the middle of a revolution." is the first sentence of *Truth, Lies and ETs*. I wrote it in March 2020, when UFOs were somewhere in the back of some people's minds. I am now writing in September 2021, when two things are on everybody's mind: covid-19 and UFOs. This book is about the UFO revolution that flared up, as unexpectedly as covid-19, while I was writing about it.

We have reached what Malcom Gladwell described in 2000 as *The Tipping Point* (99). Gladwell explained that the rise and fall of disease, crime, styles, social innovations and beliefs depends on actions and events that contribute incrementally, but individually almost un-noticeably, to change our human environment until the change suddenly becomes too big to ignore. We have reached and passed the tipping point about UFOs. Gladwell emphasized the vital role in promoting those changes of people he called Mavens, Salesmen and Connectors. Who are the Mavens, Salesmen and Connectors that created and then expanded the UFO revolution?

The Mavens are hundreds of people, past and present, in the world-wide UFO community. Mavens have been studying everything about ETs and ETVs since they first attracted widespread attention in the mid-1940s. Mavens organized themselves into private research groups, many of which have been mentioned here. They meet regularly and publish observations and opinions in books, bulletins and online posts. Mavens are people whose intelligence and curiosity outweighs their concern about other people's opinions. Some are specialists with advanced degrees but most are not. Their collective work has provided the facts about ETs and ETVs. This book is full of references to their work.

The Salesmen are the people whose print and video appearances have helped to convince everybody else that the "UFO phenomenon" is real. Salesmen who came into prominence in 2017 were the US Navy pilots who were interviewed about what they saw while chasing ETVs that were first recorded on ship-based advanced radar systems, then seen by the pilots who were sent up to chase them, and recorded again on the airborne radars in their jets. They have been mentioned here. Other salesmen were the government officials who, in 2017, spoke out about the Pentagon's interest in UFOs. They too have been mentioned. Credit is also due to a few politicians like the late Senate Majority Leader Harry Reid of New Mexico who, along with one or two of the better-placed Mavens, worked behind the scenes in the US government to finance government- funded "UFO research."

Salesmen, both pilots and non-pilots, have been reporting what they saw ever since the mid-1940s. But until 2017 there had never been as consistent, widely-reported and officially credible a set of salesmen as the men and women who chased ETVs in their US Navy jets, and the government officials who went on the record to support them.

Everyone in the western world is potentially connected by the internet, so now the people who Gladwell called "Connectors" are the people who, facilitated by the print and online media that host them, have gained the widest attention while reporting the ETV story. The connectors who made the first big impression in this narrative were Helene Cooper, Ralph Blumenthal and Leslie Kean, authors of the December 16, 2017, New York Times story, "Glowing auras and 'Black Money': The Pentagon's Mysterious U.F.O. Program," published in print and online. Many UFO stories had appeared in newspapers, magazines and on TV before, but none received as much attention as that one. That story made the Navy videos widely known and it led to media coverage of the "salesmen:" the pilots themselves.

A second push over the UFO "tipping point" came on May 10, 2021, when the *New Yorker* magazine published an article by Gideon Lewis-Kraus titled "How the Pentagon started taking U.F.O.s seriously". The article was structured around a profile of Leslie Kean, one of the authors of the 2017 New York Times article (100). In 2010, Kean wrote *UFOs: Generals, Pilots and Government Officials go on the Record* (101). The *New Yorker* article led to an immediate surge in print and internet coverage. There were more pilot interviews, interviews with other salesmen and communicators and so much new public commentary that now a day seldom passes without a new UFO-related story printed and posted on the internet.

Now that we are over the UFO tipping point: what happens next? In *The Black Swan*, Nassim Nicholas Taleb tells us that we have no way of knowing what happens next (102). Taleb's book warns us that predictions based on past events are often inaccurate because of the effect of what he described in his book's subtitle: *The impact of the highly improbable*. Although we know that ETVs and ETs are here, and we know that they have taken a direct interest in our future by abducting people and interfering with our reproductive processes, we don't know how they will behave towards us in the future. All we know now is that they are here, that they have been here for quite a while, and, to put it plainly, they are messing with us.

How ETs behave towards us in the future will depend in part at least on how we behave towards ETs now. That depends on the balance of power between ETs and humans. That balance of power has already been described: ET antigravity technology that we must master and ET telepathic control that we must defeat. I do not offer solutions; I offer a warning. We must treat the ET presence as a challenge to the continuity and stability of human life on earth. We can maintain the continuity and the stability of our life on earth if and when we succeed in convincing ETs to treat us as equals both here on earth and in the universe

Notes

1 . Donderi, D. C. (2013) UFOs, ETs and Alien Abductions: a Scientist Looks at the Evidence. Hampton Roads Press.

2. . https://www.nytimes.com/2017/12/16/us/politics/unidentified-flying object-navy.html

3 . Office of the Director of National Intelligence (2021) Preliminary Assessment: Unidentified Aerial Phenomena. \

4 . https://nymag.com/intelligencer/2019/12/tic-tac-ufo-video-q-and-with-navy-pilot-chad-underwood.html

5 · https://www.nytimes.com/2019/05/26/us/politics/ufo-sightings-navy-pilots.html

6 . Mellon, Christopher (2019) The military keeps encountering UFOs. Why doesn't the Pentagon care? Washington Post, March 9, 2018

7 . Lacataski, J. T., Kelleher, C. A. & Knapp, G. (2021) Skinwalkers at the Pentagon. Henderson, NV: RTMA LLC.

8 . Blumenthal, R. & Kean, L. (2020) No Longer in shadows, Pentagon's UFO unit will make some findings public. New York Times, July 23, 2020.

9 . Clark, J. (2018) The UFO Encyclopedia, 3rd edition, Vol 1, 501 – 504.Foo fighters

10 . Clark, J. op cit 523 – 531. Ghost Rockets

11 . Clark, J. op cit 169 – 172. Arnold

12 . Condon, Edward U (1969) Final report of the scientific study of unidentified flying objects New York: Bantam Books.

13
Clark, J. op cit 1250-1255.

14 McDonald, J. E. Science in Default: Twenty-two years of inadequate UFO investigations. In Sagan, C. & Page, T. UFOs – A Scientific Debate. Cornell University Press, 1972, pp 52 – 122.

15 . Fuller, J. G. (1966) Incident at Exeter. New York: G. P. Putnam's Sons.

16 . Clark, J. (2018) The UFO Encyclopedia, 3rd edition, Vol 1 , 309 – 312.

17 . Hynek, J. A., Imbrogno, P. J. & Pratt, B. (1998) Night Siege: The Hudson valley UFO sightings, 2nd edition. St. Paul, MN: Llewellyn Publications

18 . de Brouwer, Wilfred (2010) The UAP Wave over Belgium. In Kean, Leslie: UFOs: Generals, pilots and government officials go on the record. New York: Harmony Books

19 . Jasek, M (2000). Giant UFO in the Yukon Territory. Delta, BC, UFOBC

20 . Schulze, G & Powell, R. (2008). Stephenville Lights: A comprehensive radar and witness report study regarding the events of January 8, 2008. Fort Collins, CO: Mutual UFO Network.

21 . Brown, T. T. (1928) A method of & an apparatus or machine for producing force or motion. British Patent # 300,311 (Nov. 15, 1928).

22 . Bahder, T. B. & Fazi, C (2002) Force on an asymmetric capacitor. Army Research Laboratory, Adelphi MD,. Asymmetric Capacitor Force _v51_ARL-TR.nb

23 . Hayasaka, H. & Takeuchi, S (1989) Anomalous weight reduction on a gyroscope's right rotations around the vertical axis on the earth. Physical Review Letters 63(25), 18 December 1989

24 . Lorincz, I., Tajmar, M. et al (2013) Extended weight measurements of uncharged and charged spinning gyroscopes in the earth's gravitational field. American Institute of Aeronautics and Astronautics https://doi.org/10.2514/6.2013-3766

25 . Brady, D. A., White, H. G, March, P. et al (2014): Anomalous thrust production from an RF test device measured on a low-thrust torsion pendulum. AIAA/ASME/SAE/ASEE Joint Propulsion Conference. https://doi.org/10.2514/6.2014-4029

26 . https://www.uspto.gov/patents/basics

27 . McCandlish, M. (2017) The secret space program and the feasibility of interstellar travel. MUFON symposium proceedings, 161-188. MUFON Inc., Newport Beach, CA

28 . Randle, K. D & Schmitt, Donald R (1991) UFO Crash at Roswell. New York: Avon Books.

29 . Hall, Richard H. (2001) The UFO Evidence: Volume 2. Lanham, MD: Scarecrow Press, pp 177 – 182

30 . National Investigations Committee on Aerial Phenomena NICAP (1967): Case report on a humanoid sighting in Ririe, ID, November 16, 1967 (Ms)

31 . Donderi, D. C. (1968) Comments on the NICAP occupant cases, November 11,1968 (Ms)

32 . Hopkins, B. (2009) Art, Life and UFOs: a memoir. San Antonio, TX. Anomalist Books, pp. 218 – 238

33 . Bloecher, T. (1976) The Stonehenge Incidents – January 1974. CUFOS Conference, Lincolnwood, Il., 1976

34 . https://ufocongress.com/emily-trim; https://thephenomenonfilm.com/

35 . Siddick, S. & Nickerson, R. (2018) Exploring the Ariel School Event of 1994. MUFON symposium proceedings, 124 – 138. MUFON Inc., Newport Beach, CA.

36 . Pritchard, A., Pritchard, D. E., Mack, J. E., Kasey, P. & Yapp, C. (eds) (1994) Alien Discussions: Proceedings of the Alien study conference held at MIT. Cambridge, MA: North Cambridge Press

37 . Hall, R. H. (2001) The UFO Evidence, Volume 2, op cit

38 . Fuller, John G. (1966) The interrupted Journey: two lost hours aboard a 'flying saucer". New York: Dial Press

39 . Friedman, Stanton T. & Marden, Kathleen (2007) Captured: The Betty and Barney Hill UFO Experience. Franklin Lakes, NJ: New Page Books

40 . Webb, Walter N (1994) Encounter at Buff Ledge: A UFO case history .Chicago, IL: J. Allen Hynek Center for UFO Studies

41 . Hopkins, Budd (1996) Witnessed: The true story of the Brooklyn Bridge abductions. New York: Pocket Books

42 . Jacobs, David M. (1975). The UFO Controversy in America Bergenfield, NJ: New American Library

43 . Jacobs, David M. (2015) Walking among us: The Alien plan to control humanity. San Francisco, CA: Disinformation Books

44 . Huxley, Aldous (1932) Brave New World London: Chatto and Windus

45 . Hopkins, B. (1981) Missing Time New York, NY: Ballantyne Books

46 . Hopkins, Budd (1987) Intruders: The incredible visitations at Copley Woods. New York: Random House

47 . Freud, S. (1924) A general introduction to psychoanalysis. New York: Boni and Liveright

48 . Caron, Pierre et St-Germain, Marc (2016) Les enfants de Sylvie P. La Minerva, QC : Éditions les mystères d'Éleusis

49 . Fund for UFO Research, Inc (FUFOR). 1985 Final report on the psychological testing of UFO 'abductees'

50 . Rodeghier, M., Goodpaster, J. & Blatterbauer, S. (1991) Psychosocial
characteristics of Abductees: Results from the CUFOS
abduction project. Journal of UFO Studies, New Series, Vol 3,
1991, 59 - 90

51 . Davis, T., Donderi, D. C. & Hopkins, B (2013)
The UFO Abduction Syndrome. Journal of Scientific Exploration, 27
(1), 21- 38

52 . Donderi, D. C. (2018) The American Personality Inventory:
A test to evaluate abduction reports. MUFON symposium
proceedings, 2018: 112 – 122. Newport Beach, CA: MUFON

53 . Appelle, S., Donderi, D. C., Bellissimo, J. & Hopkins, B. (2009).
Common symbols are remembered by people self-reporting alien
abductions. Association for Psychological Science, San
Francisco, CA, May 24, 2009 (poster).

54 . Lagrandeur, T., Donderi, D. C. Appelle, S. & Hopkins, B. (2008).\
Self-reported alien abductees   remember consistent sets of symbols.
Association for Psychological Science, Chicago, IL May 13, 2008
(poster).

55 . The Roper Organization (1992) Unusual personal experiences.
An analysis of the data from three national surveys conducted by the
Roper Organization. Las Vegas, NV: Bigelow Holding
Corporation.

56 . Rhine, J. B. & Pratt, J. G. (1957) Parapsychology : Frontier Science
of the Mind. Springfield, IL: Charles C. Thomas

57 .
Rhine, J. B. (1964) Extra-sensory perception. Boston: Branden Press

58 . cited in Crumbach , James C. (2009) A scientific critique of
parapsychology. Chapter 3, pp 61 – 62 of Schmeidler, G., ed.
Extrasensory Perception, New Brunswick, NJ. Transaction
Publishers

59 . Jacobsen, Annie (2017) Phenomena: The secret history of the U.S.
government's investigations into extrasensory perception and
psychokinesis. Boston, MA: Little, Brown

60 . Eisenberg, H. & Donderi, D. C. (1979) Telepathic transmission of
emotional information in humans. Journal of Psychology, 103, 19
– 43

61 . Grau, C., Ginhoux, R, Riera, A. et al. (2014) Conscious brain-to-brain
communication in humans using non-invasive technologies.
PLos ONE 9(8) : e105225. doi : 10.1371/journal.pone.0105225

62 . Kubanek, J., Brown, J., Ye, Patrick, Pauly, K. B., Moore, T. &
Newsome, W. (2020) Remote, brain-region specific control of
choice behavior with ultrasonic waves.

63 . http://www.stopabductions.com/intelligence.htm

64 . https://www.washingtonpost.com/news/checkpoint/017/12/18/former-navy-pilot describes-encounter-with-ufo-studied-by-secret-pentagon-program/

65 . https://www.baltimoresun.com/news/bs-xpm-1997-06-25-1997176087-story.html

66 . Ed Ray, Bus driver during kidnapping, dies at 91.
https://www.nytimes.com/2012/05/19/us/

67 . Pratt, j. G., Rhine, J.B., Smith, B.M., Stuart, C. E. & Greenwood, J. A. (1940): Extra-Sensory perception after sixty years. New York: Henry Holt

68 . Festinger, L. (1957) A theory of cognitive dissonance
Redwood City, CA: Stanford University Press.

69 . www.cnes-geipan.fr

70 . Salas, R. & Klotz, J. (2005) Faded Giant. Book Surge LLC, www.booksurge.com

71 . Thanks to Jan Aldrich, documents about the Air Force UFO projects obtained from Freedom of Information Act (FOIA) requests are available online at www.project1947.com

72 . Ruppelt, E. J. (1956) The Report on Unidentified Flying Object New York, NY: Ace Books

73 . Hynek, J. Allen (1972) The UFO Experience: A Scientific Inquiry. Chicago, IL: Henry Regenery

74 . CIA memorandum, unsigned, 19th August 1952
http://www.cufon.org/cufon/cia-52-1.htm

75 . Discussion of the Robertson panel is found in Hynek, J. A. op cit, pp 168 – 169, Swords M. & Powell, R. (2012): UFOs and Government: A scientific inquiry, 138 – 203 and Clark, J. Vol 2, op cit pp 1012 – 1018

76 . James, William (1890) Principles of Psychology, Vol 2. New York: Henry Holt, 1890

77 . Kuhn, Thomas (1962) The Structure of Scientific Revolutions, 2nd edition. Chicago, IL. University of Chicago Press

78 . Battelle Memorial Institute, Special Report No, 14: Analysis of reports of unidentified aerial objects, 5 May 1955.
http://bluebookarchive.org

79 Books by Donald H. Menzel and Philip J. Klass:
. Menzel: Flying Saucers, 1953,
The World of Flying Saucers, 1963,
The UFO Enigma, 1977. Klass: UFOs: identified, 1968

Klass: UFOs: Explained, 1975;

UFOs: The public deceived, 1983,
UFO abductions: A dangerous game, 1988,
The crashed-saucer coverup, 1993,
Bringing UFOs down to earth, 1997
The real Roswell crashed-saucer coverup, 1997,

80 . Shostak, Seth (2008) Confessions of an Alien Hunter. Washington, D.
      C.: National Geographic Society
81 . Csanaday, A (2015) MIT Researcher is sure there are alien life forms
      out there, not sure why so many people think they've been here:
      National Post (June 24, 2015): MI
82 . Loeb, A. (2021) Some 'Defenders' of science might actually be
      hurting it   https://medium.com/on-the-trail-of-the-saucers/
      avi-loeb-sciencef3ed3f14f26-cf3ed3f14f26Zabel
83 . https://en.wikipedia.org/wiki/List_of_UFO_organizations

84 . Clark, J. (2018) Volume 2, op cit, pp 1190 - 1203

85 . Committee on Science and Astronautics, U.S. House of
      Representatives, Symposium on Unidentified Flying Objects,
      July 29, 1968
86 . Sagan, C. & Page, T: UFOs – A scientific debate.
      Cornell University Press, 1972, pp. 52 – 122.
87 . Appelle, S. (1990) UFOs and the Scientific Method. Science, issue
      5379, pp. 919, DOI: 10.1126/science.281.5379919b
88 . Warren, D. (1970) Status inconsistency theory and flying saucer
      sightings. Science, 170, 3958, 599 -603
89 . http://www.defense.gov/faq/pis/16.html

90 . Wells, H. G (1898) The war of the worlds. London, William
      Heinemann
91 . https://en.wickipedia.org/w/index.php?title=List_of_explanetary_
      host stars & old id=982952741
92   Hill, P. (1995) Unconventional Flying Objects: A Scientific Analysis.
      Charlottesville, VA: Hampton Roads Press
93   https://en.wikipeda.org/wiki/Velostat

94 . https://www.un.org/en/universal-declaration-human-rights/

95 . Saad, Lydia (2019) Americans skeptical of UFOs, but say
      government knows more. Gallup

https:/news.gallup.com/poll/266441/americans-skepticalufos-say-government-knows.aspx

96 . Hobbes, Thomas (1651) Leviathan, or the matter, forme and power of a commonwealth, ecclesiastical and civil. Oxford University Press

97 . Wright, Dan (2019) The CIA UFO Papers. MUFON/Red Wheel /Weiser LLC, Newburyport, MA.

98 . Diamond, Jared (2011) Collapse: How societies choose to fail or succeed. New York: Penguin

99 . Orwell, George (1944) London letter to Partisan Review, p. 339 in The Collected Essays, Journalism and letters of George Orwell, Volume 3, As I Please. Harmondsworth, Middlesex, England, 1970

100 . Gladwell, M. (2002) The tipping point. Boston: Little, Brown

101 . Lewis-Kraus, Gideon (2021) How the Pentagon began taking U.F.Os. seriously .New Yorker, May 10, 2021

102 . Kean, Leslie (2010) UFOs: Generals, Pilots and Government Officials go on the Record. New York: Harmony Books.

103 . Taleb, Nassim Nicholas (2010) The black swan, 2nd ed. New York: Random House

Picture Credits

Figure	Source
1	CBS
2	US government: public domain
3	US government: public domain
4	Creative Commons
5	Imagen
6	Associated Press
7	Wikimedia Commons
8	Wikimedia Commons
9	Wikimedia Commons
10	PARS international
11	Cornell University Press
12	Imagen
13	Paul Imbrogno
14	SOBEPS, Belgium
15	Martin Jasek
16	Martin Jasek
17	MUFON
18	UK Intellectual Property Office
19	American Physical Society
20	US patent office
21	US patent office
22	http://jnaudin.free.fr/lifters/act/html/sfptv1.htm
23	US patent office
24	Valone/American Institute of Aeronautics and Astronautics

25 . Donald Schmitt

26 . Creative Commons

27 . Marc Rodeghier

28 . Village Voice

29 . Emily Trim

30 . Kathleen Marden

31 . Kathleen Marden

32 . Kathleen Marden

33 . Kathleen Marden

34 . Marc Rodeghier

35 . Marc Rodeghier

36 . Marc Rodeghier

37 . Mary Evans Picture Library

38 . Simon & Schuster

39 . Simon & Schuster

40 . Simon & Schuster

41 . Society for Scientific Exploration

42 . MUFON

43 . Don C. Donderi

44 . US patent office

45 . Small Business Innovation Research (SBIR), US govt.

46 . Amazon Inc.

47 . US patent office

48 . Chicago Sun

49 . Roswell Daily Record

50 . Ruppelt, Almy; Hynek, U. of Chicago

51 . PARS international

52 . Menzel, CCC Marketplace; Klass, Wikimedia commons

53 . Shostak, SETI Institute; Seager, Justin Knight

54 . Lotem Loeb

55 . Don C. Donderi

56 . Encyclopedia Britannica

57 . Wikimedia Commons

58 . Getty Images

59 . ET, Almy; Spok, Wikimedia Commons

60 . New Yorker; Mike Stevens

61 . Simon & Schuster

62 . David E. Pritchard

www.ingramcontent.com/pod-product-compliance
Lightning Source LLC
Chambersburg PA
CBHW021927190326
41519CB00009B/930